QH 445.2 .S85 2002
50554275
The common thread

P9-CAO-233

THE RICHARD STOCKTON COLLEGE
OF NEW JERSEY LIBRARY
POMONA, NEW JERSEY 08240-0195

THE
COMMON
THREAD

THE COMMON THREAD

**A Story of Science, Politics, Ethics,
and the
Human Genome**

JOHN SULSTON
GEORGINA FERRY

**The Joseph Henry Press
Washington, DC**

THE RICHARD STOCKTON COLLEGE
OF NEW JERSEY LIBRARY
POMONA, NEW JERSEY 08240-0195

Joseph Henry Press • 500 Fifth Street, N.W. • Washington, D.C. 20001

The Joseph Henry Press, an imprint of the National Academies Press, was created with the goal of making books on science, technology, and health more widely available to professionals and the public. Joseph Henry was one of the founders of the National Academy of Sciences and a leader in early American science.

Any opinions, findings, conclusions, or recommendations expressed in this volume are those of the author and do not necessarily reflect the views of the National Academy of Sciences or its affiliated institutions.

Copyright 2002 by John Sulston and Georgina Ferry. All rights reserved.

Printed in the United States of America.

Library of Congress Cataloging-in-Publication Data

Sulston, John.
  The common thread : a story of science, politics, ethics, and the human genome / John Sulston, Georgina Ferry.
     p. cm.
Previously published: London ; New York : Bantam, 2002.
Includes bibliographical references and index.
  ISBN 0-309-08409-1
  1. Human Genome Project. I. Ferry, Georgina. II. Title.
  QH445.2 .S85 2002b
  611'.01816–dc21

2002014007

ILLUSTRATION ACKNOWLEDGEMENTS

Page 1. (3) Bob Horvitz: *Medical Research Council.* (4) Judith Kimble: *Medical Research Council.* (5) Max Perutz hands the key of the LMB to Sydney Brenner: *Medical Research Council.* **Pages 2/3.** (1) Dideoxy method of sequencing: *Wellcome Library, London.* (3) JS in Room 6024: *Bob Waterston.* (6) Cartoon by Bill Sanderson: *Wellcome Library, London.* **Pages 4/5.** (1) Sydney Brenner and Jim Watson: *Cold Spring Harbor Laboratory.* (2) Sequencing room at the Sanger Centre: *Wellcome Library, London.* (3) Sequence output: *Wellcome Library, London.* (5) Bermuda overhead: *Wellcome Library, London.* **Pages 6/7.** (1) Craig Venter, President Bill Clinton and Francis Collins, Washington, 26 June 2000: *Associated Press.* (2) JS, Michael Morgan and Mike Dexter, London, 26 June 2000: *Associated Press.* (3) Tony Blair, London, 26 June 2000: *Associated Press.* (4) Richard Gibbs, Evan Eichler, Francis Collins and Eric Lander, Philadelphia, October, 2000: *NHGRI/Kris Wetterstrand.* (5) *Nature* cover, 15 February 2001: *reprinted by permission from* Nature *vol. 409, no 6822,* © *Macmillan Magazines Ltd.* (6) Craig Venter, Francis Collins and Eric Lander, Washington, 12 February 2001: *Associated Press.* **Page 8.** (1) Sanger Centre panto, 2000: *Richard Summers.* (2) JS and Allan Bradley: *Wellcome Library, London.* (3) Mike Stratton: *Wellcome Library, London.* (4) JS and Marc Quinn: *Press Association/Johnny Green.*

CONTENTS

To the International Human Genome Sequencing Consortium

PREFACE

THIS IS THE STORY OF AN EXTRAORDINARY ENTERPRISE, ONE OF THE
notable achievements of late twentieth-century science: the sequencing
of the human genome. It is a story that has been told and retold in the
pages of the popular press, often accompanied by breathless headlines
and bold claims about the end of disease. And as if it were not exciting enough already, the story gained even greater appeal when a
challenger entered the lists and turned a scientific quest into a "race."

So why tell this story again? It seemed to us that only an insider's
view could do justice to the dramatic developments of the past decade
and more, which have been far too complex to reduce to the convenient
but misleading metaphor of a "race" for the genome. And, as the head
of the largest genome sequencing center outside the United States,
John was in a unique position to comment on the politics of a scientific
advance equally critical to corporate wealth and human health.

The sequencing of the human genome is the latest step in a
process that began during the 1980s, when two separate lines of
research began to converge. One was human genetics, the study
of patterns of inheritance that can reveal genetic causes of disease;
the other was molecular biology, which studies the stuff of which

genes are made: DNA, the molecule that makes life possible. Using a simple alphabet of only four letters, DNA encodes the instructions to make every living thing as sequences of those letters—the genes.

In 1990 an international, publicly funded effort, the Human Genome Project, was launched to map and sequence human DNA and make the information freely available to the scientific community. Confusingly, people often talk about 'mapping' the genome when they mean 'sequencing'. The difference is on one level a question of scale—you can make a useful map that helps you to find genes without knowing the full sequence of letters (3 billion in the case of the human genome). But although the sequence is in one sense the ultimate map, it is also much more than that: it is also the biological information itself. When we finish collecting the sequence we will have the hieroglyph of biology in our hands, even if we don't at first understand it all.

Deciphering the information will take a long time, and will need every available mind on the job. And so it is essential that the sequence is available to the whole biological community. No single individual or group can credibly claim that they have the expertise to deal with it. When the commercial company that became Celera Genomics was launched in May 1998 with the stated aim of becoming "the definitive source of genomic and associated medical information," the whole future of biology came under threat. For one company was bidding for monopoly control of access to the most fundamental information about humanity, information that is—or should be—our common heritage.

To their great credit, the public bodies that were funding the Human Genome Project quickly decided not to leave the field to Celera, but to make the sequence available to everyone even faster, though initially at a lower standard of accuracy and completeness, than they had originally intended. Thus the world celebrated a "working draft" of the sequence with much fanfare in June 2000; and, though it will be some years more before the sequencing centers finish the job, today any scientist anywhere can access the sequence freely at no cost and use the information to make his or her own further discoveries. We wrote this book so that people might understand how close the world came to losing that freedom.

Insidiously, over the past few decades, the prevailing ethos in the world of science has shifted. What was once a collective enterprise, in which discoverers were acknowledged but their results freely shared, is now frequently constrained by the demands of commercial competition. Motivated by financial gain, hamstrung by sponsorship deals, or simply out of self-defense, many researchers trade their discoveries with the rest of the community only under the protection of patent law or commercial secrecy. On the other hand, there are still many who cling to the older ideals of science and have raised their voices in protest at the way things are going. The Human Genome Project provides a unique illustration of the choices faced by individual scientists and by society as a whole. We hope that this story conveys the sheer thrill of scientific discovery, but also that it provokes reflection on the awesome responsibility borne by those who hold the secret of humanity in their hands.

A word about the "voice" in which the book is written: We wrote it together, as far as possible in an equal partnership. But we agreed from the start that as we were telling John's story, it should be written in the first person, in his voice. The principal source has been John's memory, backed up by his email files that constitute a day-by-day record of the project. In addition we have interviewed many of the other protagonists, whose perspectives have been invaluable in rounding out the story and corroborating or contradicting John's memories. The result is, to the best of our ability, an accurate and unvarnished account of what really happened.

We should like to record our thanks to the following, with all of whom it was a pleasure to talk to or correspond: Bart Barrell, David Bentley, Martin Bobrow, Sydney Brenner, Murray Cairns, Francis Collins, Alan Coulson, Mike Dexter, Diana Dunstan, Richard Durbin, Eric Green, Phil Green, Mark Guyer, Bob Horvitz, Tim

Hubbard, Jon Karn, Judith Kimble, Aaron Klug, Eric Lander, Peter Little, Michael Morgan, Bridget Ogilvie, Maynard Olson, Paul Pavlidis, Colin Reese, Dai Rees, Jane Rogers, Gerry Rubin, Mike Stratton, Adrian Sulston, Ingrid Sulston, Bob Waterston, James Watson, John White, and Rick Wilson. David Bentley, Murray Cairns, Alan Coulson, Richard Durbin, Madeleine Harvey, Tim Hubbard, Aaron Klug, Paul Pavlidis, Don Powell, Jane Rogers, Daphne Sulston, Ingrid Sulston, and Bob Waterston read versions of the manuscript in draft form, and their comments were invaluable. Any deficiencies that remain are ours alone.

We thank the funding agencies who financed the U.K.'s share of the research described here: the Medical Research Council (who also provided access to material in their archives) and the Wellcome Trust. Both helped us with pictures, as did the National Human Genome Research Institute; our thanks to Richard Summers, Anne-Marie Margetson, Jenny Whiting, Sonya Brown, Annette Faux, Kris Wetterstrand, and Maggie Bartlett.

We are sorry that, in order to keep the book within manageable limits, we had to omit many individuals whom we would have liked to include by name (for without them there would be no story to tell). We especially want to thank the members of the Sanger Centre and its partners in the International Human Genome Sequencing Consortium for tolerating John and making all this come about.

The book would not have happened were it not for three people in particular. Alan Coulson has partnered John from the early days of genomics. The collaboration with Bob Waterston sequenced the worm, founded two genomics centers, and drove the concept of free release; Bob shared in the gestation of the book and worked on its contents with us. This is his story as well. Daphne Sulston, amazed by these events, was determined to see a true account of them written down, and prompted our collaboration to do so.

John Sulston and Georgina Ferry
August 2002

PROLOGUE: SYOSSET

'I JUST HEARD THE PRISON DOOR SHUT BEHIND US.'

I stood with Bob Waterston on the glaring white platform of the little station at Syosset on the Long Island Railroad, waiting for a train to New York. The sun was harsh and bright. We were on our way home from the 1989 Cold Spring Harbor symposium on the biology of the nematode worm. It was always hard for me to go to the worm meetings—to leave behind the lovely softness of an English spring and go to that land of sharp shadows and fierce contrasts. Now I find the transition less acute, but the difference is always there.

The worm meetings had taken place every two years since 1977, but this one had been special. Alan Coulson had unrolled the long scrolls on which we had mapped out the worm genome across the end wall of the Bush lecture theatre. For all three days of the meeting he was besieged by people wanting to check details and add information before they went off on another two-year stint of the real stuff of biology—discovering where and when and how the individual genes direct the production of a fully working worm.

At one point Jim Watson, co-discoverer of the double helix of

DNA and then head of the Human Genome Project, visited the map. 'You can't see it without wanting to sequence it, can you?' he remarked. Our map represented overlapping pieces of worm DNA, lined up in the right order and studded with known landmarks. To sequence it would mean reading every one of the 100 million letters, or bases, in the worm's genetic code. That would yield the ultimate map: the total of all the information needed to make a worm. In terms of knowing our way around the worm genome, it would be the difference between a school globe and a set of street maps with every house marked.

Later we sat in Jim Watson's office discussing exactly how we might do this. And we came to an agreement. Bob's lab in St. Louis and mine in Cambridge would between us sequence the first 3 million bases of the 100 million base genome during the next three years to show that we were capable of doing it. If all went well, we would then apply for the funds to finish the rest.

That much we could plan. We had no inkling that we would soon be enmeshed in the turbulent, contentious but ultimately successful project that resulted in the announcement of the draft human genome sequence in June 2000.

On that May afternoon, as we waited for the New York train under the hot sun, the end was a long way off. Nobody had sequenced more than 250,000 bases before, let alone 3 million or 100 million. Many regarded the very idea as a waste of time and resources. Yet we had committed ourselves to doing it. In the sudden quiet after the bustle of the worm meeting, the realization hit me that there wasn't any way back: we could only go forward. The clang of the prison door reverberated in my ears. It was one of the most exciting moments in my life.

1 WITH THE WORMS

IF THERE IS A SINGLE ICON ABOVE ALL OTHERS THAT ART ACQUIRED from science in the twentieth century, it is DNA. And with good reason: this molecule, as Francis Crick famously shouted to the bemused customers of the Eagle pub in March 1953, contains the secret of life. In most representations we see it as a rather stubby double helix, for they seldom show its other striking feature: it is immensely long and thin. In every cell of your body you have two meters of the stuff; if we were to draw a scaled-up picture of it with the DNA as thick as sewing thread, that cell's worth would be about 200 kilometers long.

Like the fibers of cotton, DNA molecules can stick together side by side to make a visible thread, and this makes possible a rather lovely experiment. So when the contemporary artist Marc Quinn asked me about a DNA exhibit for his show at London's White Cube gallery in 2000, I was delighted to help. He gave me a sample of his semen (Marc is renowned for using his own body fluids in work that explores the concept of the self; in 1991 he made a cast of his head using eight pints of frozen blood), and I broke open the sperm with detergent and a special chemical that softens their tough coats. Sperm are basically just packaged DNA, and the solution became very viscous as their contents were released. We transferred a little puddle of it to a tall glass tube, and gently overlaid it with pure alcohol. Then we lowered a glass rod through the alcohol to the puddle, stirred slightly, and slowly drew the rod upwards. Tiny fibers appeared and coalesced into a thread attached to the rod. We pulled it up until it reached the top of the tube, then stuck it to the rim. Marc put the tube in front of a jet black surface, and we stood back and hugged each other at the beauty of it: Marc's DNA, a web of molecules each too small to see with the naked eye, entwined into a single shining thread. The secret of his life.

You can do a similar experiment with any living tissue—even if you don't have a laboratory to hand you can get pretty good results in the kitchen using an onion as the source of tissue, and washing-up liquid, salt and vodka to extract the DNA. It will look exactly the same as human DNA, for a very good reason: from a chemical point of view it is exactly the same kind of molecule. DNA is the common thread that links every living thing with a single primeval ancestor.

But your DNA also makes you different from an onion, and from every other human being. The DNA molecule carries a code, and the instructions that dictate whether an egg or a seed grows into a human or an onion are written in that code. Much smaller differences in these coded instructions determine the infinite variety of hair color, facial features, body shape and personality that make each of us a unique individual. Each instruction, or gene, has a small part to play in making the whole, and the overall outcome is determined in part by the environment, but the combined power of the information contained in the whole genome, the entire complement of an organism's DNA, is truly awesome. The project that is now under way to harness that power through reading and understanding the complete set of instructions that makes a human being—the human genome—is one of the most momentous enterprises in modern science. It could transform our lives, for better or worse depending on how we apply the knowledge.

Everyone seems to understand this, if the razzmatazz that greeted the June 2000 announcement that the draft human genome was complete was anything to go by. But despite the fanfares, the job isn't remotely over yet. The reading process will be largely complete some time during 2003, but the understanding will take decades and will encompass all of biology. And the generation that really understands the human genome, or the onion genome for that matter, will understand life.

I never meant to get involved in the three-ring circus of the Human Genome Project. Only ten years ago I would have laughed if anyone had suggested I would soon be directing a research center with a staff of 500, plunging into the politics of an international project and engaging in a war of words in the press. What I wanted to do was to read the genetic code of the nematode worm. I didn't imagine that the worm was going to lead us directly to the human genome. Certainly, reading worm DNA is a good preparation for reading the DNA of any other species, and that includes humans; but when we started to read the worm genome we had no thoughts of other species. We simply wanted to fill in the background to the ever more elaborate picture of the biology of this tiny creature that had developed over the previous twenty-five years.

I first met the worm in 1969, when I arrived at the Medical Research Council's Laboratory of Molecular Biology in Cambridge—universally known as the LMB—to work as a staff member in Sydney Brenner's group. Sydney was joint head, with Francis Crick, of the cell biology division at the laboratory. Physically they were a study in contrasts—Francis was tall and sandy-haired, while Sydney was short and dark with penetrating, deep-set eyes beneath startlingly bushy eyebrows—but both were great talkers. Born and educated in South Africa, Sydney had come to Oxford as a graduate student in 1952, with a medical degree but determined to work on the biology of the gene. He had quickly established himself among the international group of scientists working on the genetics of bacteriophage—tiny viruses that infect bacteria—who together were laying the foundations of modern molecular biology.

In 1953 Francis Crick and Jim Watson had discovered the double helix structure of DNA, and Sydney had been one of the first to visit Cambridge and hear about the discovery at first hand. He had moved to Cambridge permanently in 1957 and had worked with Francis on deciphering the genetic code and understanding how cells translate it into the protein molecules they need to carry out their functions. By the mid-1960s Sydney considered the work of understanding how genes make proteins almost done, and wanted to move on to the next stage. His ambitious plan was nothing less than to understand how a complete animal was encoded by its DNA. Naturally he wanted to start with something simple, and the animal he chose was the nematode worm. 'We propose to identify every cell in the worm and trace lineages,' wrote Sydney in his bid for support for the project. 'We shall also investigate the constancy of development and study its genetic control by looking for mutants.' Sydney later recalled that some people thought the idea was crazy. 'Jim Watson said at the time that he wouldn't give me a penny to do it,' he said. 'He said I was twenty years ahead of my time.'

Why did Sydney pick a worm? There is a long tradition in biology of studying simple organisms in order to discover mechanisms that are at work in all living creatures. At the time Sydney embarked on his project, most geneticists worked with bacteria or the fruit fly *Drosophila melanogaster*. But neither of these suited Sydney's purpose. Bacteria are single-celled organisms; one of the main objects of his program was to look at how the genes control the successive cell divisions that turn an egg into an adult in a multi-cellular animal. The fly, on the other hand, with its compound eyes,

wings, jointed legs and elaborate behavior patterns, was too complicated to be susceptible to the sort of exhaustive analysis Sydney had
in mind. Nematode worms, or roundworms, were not as well
studied as either, but they were far from unknown to biology. They
constitute a large family that includes both parasitic and free-living
varieties. The species that interested Sydney was a free-living soil-
dweller, *Caenorhabditis elegans*: a long name for a tiny creature only
a millimeter from nose to tail.

In the wild, *C. elegans* lives in soil and feeds voraciously on any
bacteria or other micro-organisms it can find. It grows from egg to
adult in three days (one-third of the time for a fruit fly), except when
food is scarce, when it can hang about in a non-breeding larval form
for several months. Most adults are hermaphrodites and produce
several hundred offspring through self-fertilization. Males arise
occasionally, perhaps at a rate of one in a few hundred, and mating
provides the possibility for genetic mixing which allows for more
rapid evolution. The worm's anatomy is quite simple, but although
it lacks many of the physiological features of higher animals, such as
a heart, lungs and bones, it can still carry out many basic tasks: moving, feeding, reproducing, sensing its environment and so on. It
consists basically of two tubes, one inside the other. The outer tube
includes the skin, muscles, excretory systems and most of the nervous system; the inner tube is the gut. It moves by contracting its
dorsal and ventral muscles alternately, arching its body into a series
of S-shaped curves.

The worm is, moreover, well suited to the kind of investigation
Sydney had in mind. It is easy to keep and breed in the laboratory,
living happily in petri dishes that have been sown with lawns of
Escherichia coli bacteria. You can even keep them in suspended
animation in the freezer for years at a time, allowing you to preserve
stocks of different strains of the animal. Both larvae and adults are
transparent, so that, given a good enough microscope, you can see
not only the internal organs of living animals but even individual

cells. The adult hermaphrodite usually has exactly 959 cells, not counting the egg and sperm cells. (For comparison, a fruit fly has more cells than this in just one of its eyes, and the human body has 100 trillion.) Its genome is made up of 100 million bases divided into six segments, or chromosomes.

Sydney hoped that he would be able to establish direct links between the worm's genes and its development from egg to adult, following the classic route of geneticists, in use since the first decades of the twentieth century. With a fast-breeding species, such as a worm or a fruit fly, occasional changes arise in the DNA that make the animal look or behave abnormally. These changes are known as mutations, and the altered animals as mutants. Geneticists soon developed a variety of techniques to increase the normal mutation rate. In the 1960s there was no way to analyze the DNA directly, but by cross-breeding mutants and looking at the patterns of inheritance in later generations you could map the relative positions of the mutated genes on the chromosomes. The closer together two muta-tions lay on a chromosome, the more likely they were to be inherited together. As well as mapping the genes, Sydney hoped, through careful microscopy and biochemistry, to discover exactly what was going wrong in mutant worms at the level of cells.

Assisted by a succession of young researchers, most of them American, Sydney was initially very successful in finding mutants and mapping the affected genes along the chromosomes, confound-ing those skeptics who had said that the worm was so boring in appearance and behavior that he would never be able to distinguish the mutants from the rest. But the timescale of the whole enterprise turned out to be longer than Sydney anticipated. Genes almost always work in concert, rather than solo—only very rarely is it possible to follow a direct line through from one gene to one function. Even so, the whole thing took off in a larger way than Sydney could have predicted because his intuition led him to an animal with tremendous potential for research.

As was typical of Sydney's style—indeed, the style of the LMB as a whole—on my arrival I was given about a meter of space at the bench in a crowded lab and more or less left to get on with it. Sydney and Francis believed that keeping the lab tightly packed encouraged people to interact, and that 'desks encouraged time-wasting activities.' I found myself among a group of young researchers, astonished that we were being paid to do what we wanted to do anyway, and knowing that we had no-one to blame but ourselves if we did not succeed. I compared notes with another new arrival, amazed like myself by the pride, to the point of arrogance, that we found at the lab. 'Who do these people think they are?', I remember him saying. But gradually we realized that they had a right to be proud, and as time went on we acquired some of that pride ourselves, though personally I was convinced that I could never do well enough to live up to the past glories of the LMB.

The laboratory was then and still is one of the world's top centers for research into the molecular basis of life. This was the place, more than any other, where the field of molecular biology had been invented. Its unique ethos undoubtedly played a role in shaping my development as a scientist. It grew out of a fortunate combination of circumstances in the years after the Second World War. Many academic scientists had engaged in war-related research, and the results were spectacular: radar, high-speed computing, antibiotics, and nuclear technology all had their origins in wartime research. It dawned on the government of the day that investment in science could have a long-term payoff. Up to the end of the 1930s there was little opportunity to do research in Britain if you didn't have a university teaching appointment or a private income. But ten years later it suddenly became easier to get grants, generous ones, from government-funded bodies such as the Medical Research Council (MRC) or the Department for Scientific and Industrial Research. This sudden largesse coincided with one of the most exciting periods in the history of biology, as more and more people began to apply the methods of physics and chemistry to biological problems.

Lawrence Bragg was a physicist who headed the Cavendish Laboratory, the physics department of Cambridge University. As a young man, Bragg had pioneered the technique of X-ray crystallography that made it possible to study the three-dimensional arrangement of atoms in molecules, including biological molecules. Among his staff was a meticulous, quietly spoken Viennese émigré chemist called Max Perutz. Perutz, together with a young colleague, John Kendrew, also a chemist, was trying to decipher the structure of the blood protein hemoglobin. X-ray crystallography worked well for small molecules, but proteins contained thousands of atoms and progress was slow. Bragg, an extremely influential figure in British science, was an enthusiastic supporter of Perutz's work. In May 1947 he wrote to the Secretary of the MRC asking for the funds to establish Perutz's group 'on a more permanent basis.' Within months the MRC agreed to support a Unit for Research on the Molecular Structure of Biological Systems, with Perutz at its head.

The unit, later given the slightly snappier title of the Molecular Biology Research Unit, was basically Perutz and Kendrew. They were soon joined by two research students, Francis Crick and Hugh Huxley, both physicists returning to academic life after several years of war service. Two years later the arrival of Jim Watson, then a 22-year-old American whiz-kid geneticist, opened up a whole new field of possibilities. Perutz says that it was Watson who made them realize that physics and chemistry might not hold all the answers.

Watson's arrival had an electrifying effect on us because he made us look at our problems from a genetic point of view. He asked not just, 'What is the atomic structure of living matter?' but foremost, 'What is the structure of the gene that determines it?'

Francis took little persuading that it was more important to work on DNA, then just beginning to be recognized as the stuff of which genes are made, than on protein. Francis and Jim did little experimental work themselves, but they read, talked, argued and built models, and with the crucial help of an X-ray photograph of DNA taken by Rosalind Franklin at King's College in London (shown to Watson by her colleague Maurice Wilkins) they correctly inferred the double-helix structure of the molecule and published it in *Nature* in 1953. I was an eleven-year-old schoolboy at the time, but I remember those years as a time of huge excitement that so much was being discovered.

DNA is a long, thin molecule made up of a chain of units called nucleotides; each nucleotide carries one of the four bases adenine (A), guanine (G), cytosine (C) or thymine (T). Watson and Crick deduced that two strands of DNA wind into a double helix in which A always pairs with T and C with G. They realized that this base pairing provides the mechanism by which DNA can replicate, the fundamental requirement for the evolution of life on earth. Adopting a deliberately insouciant turn of phrase that has passed into scientific folklore, Watson and Crick ended their 1,200-word paper with the sentence: 'It has not escaped our notice that the specific pairing we have postulated immediately suggests a possible copying mechanism for the genetic material.' In other words, if you have a single strand of DNA and an unlimited supply of the four nucleotides, you can make the second strand, and from this another copy of the first strand, and so on.

A month after their first *Nature* paper came out, Watson and Crick followed up with another which spelt out a further momentous consequence of their discovery: '[I]n a long molecule many different permutations are possible, and it therefore seems likely that the precise sequence of the bases is the code which carries the genetical information.' They were right; and the practice of biological research has been changed for ever by this understanding. This is the truly remarkable outcome of knowing the structure of

DNA—not the helical form itself, but the confirmation that the system for conveying the instructions for making a life from one generation to the next is digital, not analogue—like the English language, and not at all like, for example, a blueprint. To convey the notion of a furry animal that has whiskers and purrs, a speaker of English says the three-letter word *cat*, which stands for a cat to all those who understand the language. A blueprint, on the other hand, shows a graphical representation of a cat. Cat DNA spells out genes (the instructions to make a cat) just as the alphabet spells out words in a language. There is no graphical representation—nothing like the tiny homunculus curled up in the head of a sperm which some of the earlier microscopists imagined they could see.

The human genome consists of 3 billion base pairs of DNA, parcelled out into 24 chromosomes. They are numbered from 1 to 22, plus the X and Y sex chromosomes. The nucleus of every cell in our bodies (other than the egg and sperm cells, and red blood cells) contains two sets of chromosomes, one inherited from each parent: two each of chromosomes 1 to 22, plus one X and Y in males, or two Xs in females. If you scaled up the thickness of the DNA chain to that of ordinary sewing thread, you would need a 4 kilometer reel to represent the length in an average human chromosome.

When people talk about the sequence of the human genome, they mean the order of the bases on one of the paired DNA strands from each of the twenty-four different chromosomes. It doesn't matter which strand you choose, because the sequence of one immediately gives you the sequence of the other, according to the base pairing rule. The details of the chemistry of DNA give each strand a direction (like a series of arrowheads), and, as Francis and Jim recognized, the strands run in opposite directions. Scattered along the genome, some on one strand and some on the other, are stretches of sequence, at first sight no different from the much longer intervening stretches, that are the genes. I'll go into the way genes work in more detail later on p. 39. For the most part, they instruct the cell to make proteins, which themselves consist of chains of small molecules called amino acids.

Around the same time that Jim and Francis were solving the DNA structure, a Cambridge biochemist called Fred Sanger was the first to work out the complete sequence of amino acids in a protein, insulin. Fred is a quiet, unassuming, self-contained scientist with a tremendous capacity for seeing a difficult practical problem through to its conclusion. His work on insulin proved beyond doubt that proteins were not assembled according to any simple chemical rule, and that their construction must therefore be directed by a set of instructions encoded in the genes. At that stage Fred worked in the Department of Biochemistry at Cambridge University, not Max Perutz's unit, but he too was funded by the MRC as a member of its external staff. By the end of the 1950s Perutz and Kendrew had done the seemingly impossible and discerned the three-dimensional structures, formed from elaborately folded protein chains, of both hemoglobin and its smaller cousin myoglobin—the first protein structures to be solved.

It was obvious to the MRC that in supporting molecular biology they had backed a winner. More and more people wanted to join Perutz's group, which was clearly getting too big to remain in the Cavendish Laboratory. In 1962 the Research Council opened a new, six-story building next to Addenbrooke's Hospital on the south side of Cambridge. It was called the MRC Laboratory of Molecular Biology (LMB), and Perutz was its first director. John Kendrew, Francis Crick and Sydney Brenner were all still there. Fred Sanger moved across from biochemistry to head the protein chemistry division. There was a constant floating population of short-term researchers and distinguished visiting scientists from all over the world. And the discoveries continued to flow; among them, most importantly for our story, Fred Sanger's 1975 invention of a method for reading the sequence of DNA. Altogether, nine Nobel prizewinners have made the trip to Stockholm as a result of work they did in the LMB or the unit that preceded it.

The significant point is that the MRC was prepared to finance long-term work. It took Max Perutz twenty-three years (less the war years, when he was first interned as an alien and then released to conduct war-related research) to solve the structure of hemoglobin, and many chemists and biologists thought he was wasting his time. It wasn't certain, when he began, that proteins even had a stable structure. But Lawrence Bragg supported him, because although he thought Perutz's project might not work, he knew that if it did work it would be very important. And that was the tradition that pervaded the LMB. It wasn't regarded as foolhardy to take on projects when you couldn't necessarily see how you were going to carry them out, as long as they were important enough. You didn't—and still don't—have to justify everything in advance; you were just given the time, and a limited amount of space and resources, to get on with it.

Just as important in characterizing the ethos of the LMB in its heyday is that the work was not done in pursuit of any ulterior motive, financial or otherwise. Yes, the discovery of the structure of DNA opened the way to the biotechnology era; but that's not why Jim and Francis did it. Francis himself put it best in a letter of reproof he dispatched to Jim in 1967 after reading a draft of Jim's superb account of the discovery, *The Double Helix*, in which Jim implied that his eyes were always on the Nobel Prize: 'The major motivation was to understand.'

It was an environment that suited me perfectly. I don't think I would have survived a conventional academic career, juggling teaching, research and administration in a university. I was incredibly lucky to end up where I did, as my progress as a scientist up to this point had been somewhat erratic. I was without any focused ambition, simply moving from one thing to another as friends and colleagues advised.

Studying science at school and university was no more than a natural progression from my childhood interests. Right from the beginning I had surrounded myself with construction sets such as Meccano and all kinds of electrical gadgets—I was fascinated by electricity. I made radio sets, and begged a broken-down TV from a TV shop, which I fixed so that you could just about watch it in a darkened room. I kept pond life in an aquarium, and watched *Hydra* doing its crazy handstands through an old microscope my uncle had given me. Long before I started school biology in any serious way I dissected a dead bird that I had found, and was fascinated to discover that living things were also mechanisms.

My father, an ordained minister in the Church of England, was also interested in things to do with the natural world. My mother, an English teacher, was a very pragmatic person who was happy to answer my and my sister Madeleine's questions—we would talk and talk. We were not especially well off, although my father had a slightly better income than he would have had as a vicar: he became overseas secretary of a missionary society, the Society for the Propagation of the Gospel. He was never a missionary himself; he joined the Society after having been an army chaplain in Egypt and spent the rest of his career as an administrator.

For me, adolescence meant dealing not just with sexual traumas but with the question of what to do about Christianity. I'd been brought up in the Church and was a complete believer as a young child. But in my teenage years I began to question. My science was beginning to go beyond the realm of fiddling with construction sets; I remember being excited to discover that the mind could reach out and understand things that were very large and very small, from planetary motion to the power in a grasshopper's legs. I think many scientists have felt that: a sudden overwhelming sense of the power of the human mind. It forces you to look at beliefs based on articles of faith and say, 'These don't really measure up; sorry.' Of course, it's not that the power of observation and deduction actually disproved the existence of God—there are plenty of scientists who are also believers—it was just that, as a strategy for living, religion didn't make much sense to me.

I also differed from my father in my view of human society. He had a conventional view of hierarchy and class, and this was something I was amazed by because for me the central thing about humanity is that we have to be treated equally. But although I couldn't share his views on religious belief and social class, my father's influence has always been an important factor in my own approach to life: I grew up with his indifference to material wealth, and his overriding sense that one should work for the common good. Trying to live up to his standards in this respect has been a factor in shaping my views on how we gather scientific information and make it available to others.

I had what amounted pretty much to the usual education for a middle-class boy of the 1950s. I was sent to a local private preparatory school, from where I got a scholarship to the London Merchant Taylors' School. Originally a sixteenth-century foundation, the school then occupied a range of rather grand, purpose-built 1930s brick buildings surrounded by playing fields, near where we lived in the London suburb of Rickmansworth. It offered a fairly traditional curriculum to academically able boys; I fell naturally into the science stream and went along with the general assumption that I would in due course go to Cambridge and read for a degree in natural sciences.

When I came to Cambridge as an undergraduate in 1960 I was still interested in living things. I wanted to specialize in neurophysiology (living things and electricity, all in one package!). But somehow, either because of my exaggerated expectations or because of the teaching, it didn't work out, and I ended up with organic chemistry, which was very well taught. I rationalized to myself that organic chemistry was a good basis for biology—I think rightly: it gives you a sense of molecules and what they do, and indeed it's reasonable to view biology as just a branch of chemistry.

It has to be said that I was not a model student. I started doing theatre lighting for the Amateur Dramatic Club early on, despite the disapproval of my college dean, who said anyone who got involved in theatre would fail their exams. I duly did rather poorly in my second-year exams. In the Cambridge system this doesn't matter too much, but it made me realize that if I didn't get my act together I might fail my finals as well. So I offended my friends in the theatre by backing out of the main production in my final year. I knew I just had to put my nose down and get on with the work. Determined not to fail, I set myself a rigid revision schedule and just kept up with it until the final exams. But it was a total grind: I wasn't deeply interested. I managed to get a 2:1—an upper second-class degree—but it was a bit of a near squeak.

I never meant to do a Ph.D.. It was quite clear that I had no talent or taste for this book-learning stuff, so I signed up to go abroad with Voluntary Service Overseas. Then, as it happened, my VSO scheme fell through just as I got my results. I wandered along to the chemistry labs, more or less on the rebound, and asked about becoming a research student. It was the 1960s, a time of university expansion: the doors were open, and a 2:1 was good enough to get me in. I couldn't have done it now.

I joined a group under Colin Reese, a young lecturer only four years into his first teaching job, who was working on nucleotides, the building blocks of DNA and RNA. RNA is another nucleic acid that carries the coded DNA instructions out of the nucleus and acts as a template for the assembly of proteins. Colin was interested in using chemical methods to make synthetic nucleic acids by stringing nucleotides together in a predetermined order. A lot of advances in biology, including genome sequencing, have depended on the availability of synthetic DNA and to a lesser extent RNA. But Colin worked on the problem mainly because he found it scientifically interesting. I went into this with no thoughts of solving great problems in chemistry, but after a few weeks I was completely obsessed. It was like being back in my bedroom at home with all the toys. Every day there was a practical problem to solve, and it was fun finding ways to make things work. I did what I liked, and often did it differently from the way you were supposed to, and then of course you stumble across things. I learned about everything: mass spectrometry, nuclear magnetic resonance—all the new big toys I managed to wangle my way into as research students do. I synthesized new compounds and got my name on lots of papers. Colin wrote it all up—I had nothing to do with it, I was a technician—but I had a lot of fun, learned some chemistry and got a Ph.D. at the end of it.

The question of what I should do next was solved for me more or less before I even began to think about it. Colin knew Leslie Orgel, a former Cambridge chemist who had recently moved to the Salk Institute in La Jolla, California. Leslie had started out as a theoretical chemist in Oxford, but after moving to Cambridge and hanging out with the molecular biologists in Perutz's unit, he went off in a completely new direction. He was now doing practical organic chemistry in an attempt to work on the mechanisms by which life might have begun on the primitive earth. At the Salk he was building up his group and asked Colin to recommend people, and Colin gave him my name. I guess Leslie knew that Colin's students were well trained and were likely to get something done. I was the second of them to be invited, joining a trail of young Ph.D.s going over there and having a good time in Leslie's lab, being spoiled and taken out to dinner with the great and good of science.

The year 1966 was extraordinary. As well as finishing my Ph.D. and getting a job in California, I had got married. During my second year as a research student I had met Daphne Bate, a research assistant in the geophysics department who used to come round with a friend for dinner regularly at the flat I shared with one of their colleagues. During that year Daphne and I didn't pay any particular attention to each other, but a bit later we began to realize that we wanted to be together. Within a few months we had to face the question of what was going to happen next. Was Daphne to give up her job to go with me to California? And if so, should we get married? It was quite clear that both sets of parents would be very upset if we went to America together unmarried. So eventually, after much discussion, we married in the late summer, not long before I was due to leave.

Just as we arrived in California we realized that Daphne was pregnant, which was completely improper and not planned. We thought we were going to be destitute; we knew that health care in the United States was expensive, and although general health insurance came with the job it did not cover pregnancy and childbirth. But somehow everything worked out fine. My post-doc's salary was enough for our simple tastes. Instead of living in an apartment near the La Jolla campus like the other post-docs, we rented a wooden house in an unfashionable area near the beach at Del Mar. We grew vegetables in the garden behind the house, and we saved enough money for paying the doctor's bills not to be a problem. We continued to live very happily in that house after our baby daughter Ingrid was born. We'd bought an old pickup and we used to drive to the beach with the pram in the back. Sometimes I'd walk the five miles along the coast to the lab. It was completely unspoiled and very beautiful. On our holidays we visited national parks from Canada to Yucatan in an old VW Beetle we'd bought when we realized that the pickup wasn't up to long journeys.

I sometimes describe myself as a child of the sixties, and that's all the excuse some journalists have needed to label me an 'ex-hippy.' But my experience of the sixties wasn't like that at all—it was nothing to do with rock concerts and dropping out. It was a matter of not living lavishly but enjoying what you had, growing things with your hands, working hard but not being tied to a nine-to-five job, and generally feeling that there's more to life than money. And all this was set against a background (in the U.K.) of sufficient spending on public services, which gave a great sense of security. I regret that nowadays we seem to have lost too much of that, and live in a world in which we are materially richer but apparently nothing matters except next year's bottom line.

I stayed in Leslie Orgel's lab for two and a half years. It was very different and exciting, and broadened my scientific horizons enormously. It was there that I first really understood the concept of evolution—partly, as so often happens, because I had to explain it to someone else. How is it that the chance events of environmental and genetic change can give rise to organisms that seem perfectly designed for their lifestyles? How, indeed, did life evolve from non-living matter in the first place? The sole requirements for evolution are replication and inherited variation. In other words, the evolving organism must be able to reproduce itself, must do so imperfectly, and the variations must be transmitted to the next generation. A crystal replicates itself when placed in a saturated solution of its salt, but it cannot evolve because there is no inheritance of variation.

The beginnings of evolution and the origins of life are one and the same. Once something is replicating with variation, it will bit by bit explore the possibilities of its environment. Its more successful descendants will colonize at the expense of the less successful: the process described by Charles Darwin which he called natural selection. Leslie put me on his long-standing project to investigate how the first nucleic acids might have replicated without the evolutionarily more recent enzymes that are essential to the process in all modern organisms. I didn't make a big contribution, but we obtained some results that we wrote up. Since then the lab has worked to look for analogues to nucleic acids that might have formed on the primitive earth, to see whether they could replicate more easily.

For me, the time eventually came to move on once more. There were two options open to me. There was a move to establish a middle tier of staff at the Salk, between the founding fellows and the post-docs, and Leslie proposed to put me up for one of these positions. At the same time I heard from Francis Crick, who was then a visiting fellow at the Salk, that he and Sydney Brenner were expanding the cell biology division at the LMB back in Cambridge and were looking for staff. I remember Francis coming to interview me. It probably wasn't the first time we'd met, because Francis made regular visits to the lab. We sat on lab stools at my bench, and I chatted to him about what I was doing. I didn't treat it as a formal job interview, and as far as I could tell, neither did he. But I found out later that he'd also written to Colin Reese to ask him for a reference, and the upshot was that Sydney invited me to come and join the group.

Daphne and I were faced with a difficult choice. We loved the open spaces of America, and intellectually the Salk was an exciting place to be. But accepting a permanent post there would be a big step. Daphne wanted to go back to England, and although I was more neutral my parents and sister were there and I had a certain sense of rootedness. If I had been more ambitious I would probably have stuck with Leslie, but at the time other things seemed more important, and in any case I wasn't really ready for the sort of independent position that he had in mind. We decided to accept Sydney's invitation, going to the LMB on a one-year visit, but keeping open the option of returning to the Salk afterwards. In the summer of 1969 we took a last trip, driving right across to Florida, back up to Chicago and over the border into Canada, before ending up in New York where we sold the old VW. This was America, freedom and openness! Then Daphne, Ingrid and I flew back to England.

For the first few weeks we felt quite lost. I remember being astonished at how small everything looked—the cars, the houses. But we soon got our bearings, bought a car and found a house— we'd come home with almost enough saved for the deposit—in Stapleford, a village a couple of miles south of Cambridge. Daphne was pregnant again (planned this time!) and our son Adrian was born later the same year. By the time Leslie put us on the spot when he visited a year later and asked me to decide whether to return to the Salk, we had become happily settled as a family in Cambridge, and decided to stay put. We've lived in the same village ever since, with just one move of half a mile to the house we live in now. As they grew up, Ingrid and Adrian could cycle to the local village primary and secondary schools, then to sixth form college in Cambridge itself. Daphne enthusiastically began a new career as an academic librarian.

Sydney had been working on his worms for five years when I arrived, with only two other permanent people in the group. Nichol Thomson was an electron microscopist. One of the reasons Sydney had chosen *C. elegans* for his project was that Nichol had already taken good pictures of it with the electron microscope, shaving worms into fine slices so that you could see the structure of every cell in section. Then there was Muriel Wigby who assisted Sydney with the genetics, breeding mutant strains and tracking the mutant forms through the generations in order to locate the genes. Sydney was primarily interested in the uncoordinated mutants, worms with genetic defects that disrupted the graceful sinuosity of normal worm locomotion, though he collected everything he saw, both as markers for genetic mapping and for general interest. Working with no other help at first, he found many genes that could give rise to this form of defect, and named them *unc-1*, *unc-2*, *unc-3* and so on. He hoped he would be able to look at the electron micrographs, see the changes in the anatomy of the muscles or nervous system, and correlate those with the genes and with the behavior of the worms to reveal how the whole system of locomotion was controlled. But that was only part of his plan. As Sydney wrote in his original proposal, he wanted to trace lineages. By that he meant following the sequence of cell divisions that would turn a fertilized worm egg into a fully differentiated adult. It had been known for almost a century that every individual worm was likely to develop through exactly the same sequence of steps, which is not the case in mammals or even in flies. Once he knew how development progressed normally, he could begin to ask questions about how the genes controlled that process. But it was not an easy task, and those charged with undertaking it had made only limited progress.

Like several others in the group, I originally worked on the chemistry of the worm nervous system—identifying those nerve cells, or neurons, that used particular chemical neurotransmitters to communicate with their neighbors. The idea was to find mutants that affected the production of neurotransmitters. I tried out a technique that I'd learned during one of the Salk Institute's regular summer schools, when scientists from elsewhere in the United States would come to our beautiful clifftop campus by the Pacific and teach practical courses. By then I already knew I might be working on the worm with Sydney, so it seemed a good idea to learn some neurochemistry. I learned to do a reaction that involved gassing freeze-dried tissue sections with formaldehyde. The reaction gives you a beautiful fluorescent derivative of the neurotransmitters adrenaline, noradrenaline and dopamine, so that the cells light up under the microscope. I had little idea what the point of it was, but it was a trick that I could do.

In Sydney's lab I had a go with it on the worm. To begin with I couldn't get it to work, because worm neurons are much smaller than the mammalian neurons I'd worked with before: some of the nerve fibers are less than a thousandth of a millimeter across. The problem was to get the changes in temperature right during the freeze-drying process so that the neurotransmitter molecules didn't diffuse out of their original positions. I found a way of doing this using a combination of a cold block of metal and high vacuum that worked better than much more elaborate devices. I got very beautiful pictures: the detail was quite exquisite. You could see little packets of green vesicles and plot out the patterns of the neurons. I found that the transmitter in the vesicles was dopamine, and over the next few months collected several mutants that had abnormal patterns. Playing with these mutants led me directly to the discovery of another class of mutants that lack the sense of touch, tested by stroking the worms with a single eyebrow hair. These proved to be a rich source of study for Marty Chalfie, an American post-doc who arrived in the lab later and subsequently set up his own worm lab at Columbia University in New York.

It didn't seem to matter what you did—the worm was still virgin territory, you just couldn't help finding things. At some point during the first three years, Sydney put me on to determining the quantity of DNA in the worm's genome. The technique involved making a comparison between the genome you were interested in and the genome of the bacterium *E. coli*. It both measures the size of the genome and gives an idea of how much of it is taken up with repetitive sequences that don't add anything to the information content. The answer I got was that the *C. elegans* genome was twenty times the size of *E. coli*'s, which at the time we thought meant 80 million bases (megabases). But when *E. coli* was accurately mapped, it turned out to be larger than was previously thought, so the worm estimate was revised to 100 megabases. Years later, when we finished sequencing the worm, we found that was spot on. Better than I deserved, really: some of my other measurements at the time weren't so precise. The measurements also showed that the worm genome included 15 percent of repetitive sequences—something that caused trouble when we started sequencing.

I did a few other experiments with worm DNA, and helped an American Ph.D. student, Gerry Rubin, to do some similar work with yeast. Gerry, originally from Boston, Massachusetts, was ambitious but at the same time pragmatic and cheerful. Although still a student, he was full of confidence, going out and setting up collaborations with other scientists. We were very limited in what we could achieve, because the tools for analyzing DNA in any detail had yet to be invented. But we did what we could: the important thing was that we were working with whole genomes, the complete set of instructions for each organism. By the time Gerry went back to the United States in 1974, studies of genomes were about to take off. Restriction enzymes and cloning, key tools in modern molecular biology, had just become available. That meant you could chop up a whole genome into fragments and grow clones of the fragments in bacteria; the full set of clones constituted a library of the genome. With such a library at your disposal, you could at last begin to marry up the genetic information derived from studies of mutants and breeding experiments with the physical information carried in the DNA. Gerry immediately began work on a library of the genome of the geneticists' favorite species, the fruit fly, and later became one of the world leaders in fly molecular genetics.

My own future seemed much less certain. Unlike most of the other young researchers, I was a member of the MRC staff rather than a visitor funded by a grant. But the post was not a permanent one. We had two children and a house, and I really needed some long-term security. I liked what I was doing at the LMB, but could not see how it could justify giving me a tenured position such as those held by much more established scientists who ran their own research groups. I hadn't even written any papers—the first to come out with my name on it since I arrived was the one with Gerry on yeast DNA, but that was really his work. I talked over the problem with Sydney, and then with Max Perutz, the LMB's director. Max came up with the idea of a kind of 'second-class' tenured position, and asked if I wanted to be taken on on that basis. I accepted with relief; it meant financial security for my family, without the burden of responsibility that I felt would come with being a fully fledged staff member and having to run a group.

At almost the same time I embarked on a new project which at last would both give me something to put my name to and represent a real contribution to the growing picture of worm biology that

Sydney and his colleagues were developing. While I was drawing the patterns of dopamine cells shown up by the formaldehyde staining method, I'd made other pictures using a different stain to show up all the cell nuclei, and was trying to match up the two sets of pictures to work out which nuclei belonged to neurons. I suddenly realized that when worms were first hatched they didn't have as many neurons in the ventral cord, the main nerve pathway that runs the length of the worm, as when they were older. But all the textbooks said that nematodes had their full complement of cells when they hatched from the egg. And I said, 'Look, there are fifteen ventral cord neurons when it's hatched, and fifty-seven when it's older. How?' I was intrigued. One of Sydney's original aims had been to study the lineages of cells in the worm embryo; now it seemed that the cells continued to proliferate in the larva, and the question of how they did it was up for grabs. It was something I immediately wanted to pursue.

The lab had bought a special type of microscope for studies of the embryonic lineage, called a Nomarski or differential interference contrast microscope. Not only did it magnify more than 1,000 times, it also enhanced the contrast between different regions of the specimen, so that you could see individual nuclei of cells in living tissue without having to stain them. But previous researchers had never been able to follow beyond the first few divisions of the fertilized worm egg, and eventually they gave up.

If it was so difficult in the embryo, how was I going to do it in the larvae, which had a tendency to wander off just as you had them in focus? People had previously tried to stop the worms moving by squashing them or anaesthetizing them, but the results were never any good. It turned out that there was a trick to doing it. I made little agar pads, nicely smooth at the top and not very thick, so that I could focus on the worm and illuminate it properly through the agar. I popped a worm larva onto the agar and dropped a cover slip over the worm. I had painted a layer of bacteria—worm food—in the middle of the cover slip, on the underside. The worm would browse on the bacteria in this field, and every time it got to the edge of the bacteria it would turn and come back in. And it was happy— like a cow in a field. Cows aren't moving around much because they're happily munching, and worms are the same. Now, to my amazement, I could watch the cells divide. Those Nomarski images of the worm are the most beautiful things imaginable. And they're moving—you can see the worm undulating slowly around, munching, and you can watch the nuclei at the same time. I still find it incredible.

Seeing the first cell divisions in the worm larva was an absolute revelation: just knowing it was possible was extraordinary. In one weekend I unravelled most of the postembryonic development of the ventral cord, just by watching. I found out exactly where the extra forty-two cells came from. The first ten moved into the ventral cord from outside the plane of focus; the original fifteen did not themselves divide. But the ten new cells did, each producing six descendants, one of which was not a neuron. Of the fifty new neurons, four migrated to a different region and four died, leaving forty-two, which with the original fifteen made the adult complement of fifty-seven.

I began to draw what I saw, filing my drawings in a series of green ring-binders that still sit on the shelf in my office. I made my own representations of the nuclei. Looking at the books now, you can see that as I went on I got more and more confident. It became a routine: I'd pick up a worm of the correct age and say 'I'll follow this.' Some cells I followed for days: I found that if I put the larvae in the fridge overnight the cells would stop dividing where they were, and carry on the next morning when they warmed up again.

I didn't know anything at the time about what the ventral cord cells did, other than that they were nerve cells. But I knew someone who did. Working in Sydney's group was John White, who had joined at the same time as me. Because molecular biology was a relatively new field, most of the researchers at the LMB had trained in other disciplines, and John was no exception: before coming to the lab he had worked as an instrumentation engineer. Now he was in charge of reconstructing the anatomy of the worm nervous system. Nichol Thomson, slicing up the worm like salami, had produced a complete set of fixed and stained sections which had all been photographed with the electron microscope. Sydney hired John initially to develop a computerized system for tracking and recording the nerve cells through the 20,000 or so electron micrographs. He bought a computer and John got to work designing a device to view the pictures. But he gradually realized that the storage and processing capacities of the computer, even though it was quite an advanced model for its time, were inadequate to the task. And because in many respects the worm was so simple—the ventral cord was just a bundle of parallel fibers, for example—it turned out to be just as efficient to do it by eye.

John is a wonderful scientist, a sophisticated engineer who is always willing to fall back on simple methods if necessary. He abandoned the computer and gave the job to a technician in the division, Eileen Southgate. 'She would go through the successive sections marking processes she saw as equivalent,' he says. 'Then we'd get together and sort out anomalies.' This worked splendidly, and in time they were able to draw the complete wiring diagram of the worm's nervous system. Eventually published in 1986, *The Mind of the Worm* described all 302 nerve cells and the 8,000 connections between them—a magnificent achievement. Meanwhile John invented many devices for worm microscopy, but his big commercial success was the first practical confocal microscope which he and others in the LMB developed in the late 1980s. By shining an illuminating beam on a narrow point, focusing the detector on the same point and scanning to and fro, it enabled you to look at whole samples of tissue, rather than thin sections, focusing at different depths throughout the sample.

Although the computer didn't serve the worm's nervous system very well, it did change Sydney's direction. He disappeared into its electronic bowels and wrote his own operating system. The lab became a dangerous place for those of us who liked to be in around midnight. If caught by Sydney, one would find oneself listening to the latest twists of the cybernetic plot until the small hours. This was very strange to me at the time, but years later I fell prey to the same bug when we started the map of the worm genome, and throughout the mid-1980s I programmed obsessively.

At the time I was working out the ventral cord lineage, John had just about completed his analysis of the ventral cord anatomy. I burst in to see him on the Monday morning after my weekend's work (he'd been off sailing, so I couldn't get hold of him any sooner) and showed him my pictures. He was soon able to see that the seven different types of motor neuron (the nerve cells that control the worm's movement) he'd identified in the ventral cord each had characteristic lineage histories. It was an early example of Sydney's grand plan coming into effect: working on quite separate projects, John and I had fortuitously arrived at a common understanding of a small part of the worm's biology. As the work went on, the excitement affected the whole lab. At one point Sydney bet John a bottle of wine that a particular cell would have a particular history, and lost. He handed over the bottle at a group meeting in the seminar room, and, not having a corkscrew to hand, John decided to open it by injecting pressurized freon gas through a hypodermic needle. The resulting fountain of red wine left a stain on the ceiling for many years afterwards, a reminder of the euphoria of that time.

Despite the success of my first study, it wasn't immediately obvious to me that it was worth doing the whole lineage. But in the autumn of 1974 there was a new arrival in Sydney's lab. Bob Horvitz originally hailed from Chicago and took degrees in math and economics at the Massachusetts Institute of Technology before going to Harvard to do a Ph.D. in biology. Bespectacled, intense and

extremely thorough in his approach, he arrived at the LMB steeped
in the high-tech ambience that he had absorbed as a research student
first with Jim Watson and then with another pioneering molecular
biologist, Walter Gilbert, who was developing an alternative method
of DNA sequencing to Fred Sanger's. When Bob looked at what I
had been doing, just looking down a microscope and drawing, he
was unimpressed. 'Where are the data?' he asked me. He couldn't
understand how you could do any kind of analysis if you didn't have
something like a tape of readings from a scintillation counter. I
asked him, 'What makes you more ready to believe what your eyes
see on a little piece of paper that's processed the contents of tubes
in a machine than what your eyes see when they look directly
through a microscope?' I convinced Bob that they were both data,
and that my observations were at least as rich in data as his
measurements, if not more so.

It was Bob who eventually said, 'Look—the rest of the lineage is
just waiting to be done, why don't you do it?' (It was not to be the
last time that Bob tried to get me organized.) But I said, 'No, it's too
much.' One way or another, we decided to do the larval lineage
together. Just as I had begun with the nervous system, Bob began
with muscle cells, which also multiplied in the period between
hatching and adulthood. It was the first of a number of successful
partnerships I've enjoyed in my scientific life. We never formally
divided up the lineage between us—we just started with different
parts, sometimes we worked together to solve a particular problem,
and we kept going until it was complete. The only part we didn't do
was the development of the gonad. I'd written to David Hirsh, one
of Sydney's earliest American visitors, who now had his own lab in
Boulder, Colorado, to tell him about the lineaging. I knew that his
lab was working on the development of the gonad, so I said, 'Why
don't you do your bit if you want to?' He wrote back immediately,
saying, 'Your letter blew my socks off.' He gave the project to his student Judith Kimble, so she was working on that at the same time

that Bob and I worked on the rest of the worm. We published the post-embryonic lineage in 1977, not long before Bob returned to the United States to set up his own worm lab at MIT. Judith and David published the gonad lineage a couple of years later.

I met Judith for the first time at the first of the international worm meetings, which was held at the Marine Biological Laboratory at Woods Hole in Massachusetts in 1977. These occasions are great opportunities for networking and cementing the relationships that hold a research community together. On the strength of our conversation at Woods Hole, Judith decided to come and do her post-doctoral stint with me. It was the first time I'd had anyone to supervise, but fortunately she neither needed nor wanted very much supervision. Tall and blonde, she struck me immediately as strong and independent-minded—and, just as immediately, she saw that there was no risk that I would try to control what she did.

I knew that if I went to work with John he would be a colleague rather than a boss—he wouldn't be telling me what to do. This was important to me. I did not want to post-doc with someone who would keep me from following my own ideas.

In the couple of years after we finished the post-embryonic lineage Bob Horvitz, Judith Kimble and I turned from asking what happened in the developing worm to asking why it happened. A big question in studies of development is: What determines the fate of each cell? Is it genetically preprogrammed to become a nerve cell or a muscle cell—or even to die, as we discovered that some specific cells always did in the course of development? Or does it develop its identity in response to signals from its neighbors? Because the worm was so unwavering in its progress from one fertilized egg to 959 adult cells, it provided an ideal opportunity to answer these questions. We tackled them via two methods, laser ablation and genetics. John White invented a device, using a laser, to knock out just one cell in a worm, either in a developing animal or an adult. It was a technique that appealed to me, being simple and quick. We used it to show that the invariance of worm development was not entirely down to individual cells, but sometimes depended on interactions between certain cells and their neighbors. At the same time, Bob went looking for mutant worms that had the wrong number of adult cells, in the hope of finding genes that influenced normal development. He did the breeding experiments, and I looked at the lineages in the mutant worms. By the end of the 1970s we'd found twenty-four mutants that implicated fourteen genes directly in controlling cell division, and it was clear that different genes affected different lineages. We realized almost at once that the story was going to be much more complicated than one gene controlling one cell division. It would be a matter of several genes acting in concert at each stage—and we're still right at the beginning of understanding how this works.

I enjoyed the experiments with laser ablation and lineage mutants, but I had a sense of unfinished business. We had documented all the cell divisions in the larva, but what happened before the eggs hatched was still a mystery. Many people thought it was impossible to follow all the divisions in the embryo. The results of the various attempts that had been made were little better than had been achieved at the turn of the century, for other nematodes, by researchers who looked at fixed and stained embryos and tried to put them in order. I could see that the question needed a new approach, but for a long time I hesitated.

In 1977 something came up that should have got me going on it. At the Woods Hole worm meeting I was taken aside and told that a paper was being submitted from one of the other labs describing the embryonic lineage of the gut cells. But there was a dispute; others were doubtful whether the lineage was correct, and it would be better for the worm community if this was sorted out quickly. Would I look for myself and arbitrate? Well, that was an interesting and flattering challenge. The gut cells are the largest and easiest to follow, and within a couple of sessions I could tell that the proposed lineage was incorrect. A few days later I sent a report containing the complete gut lineage to everyone concerned, and the paper was withdrawn.

But it was only in 1979 that I started looking again, casually at first. Indeed, something that may have helped was a sense of despair. I had now been at the LMB for ten years. With only a handful of papers to my name since my arrival, I had achieved little as far as I could see, and felt it was high time to move on and do something different with my life. I tentatively enquired about a couple of jobs outside research. But my explorations hadn't turned up anything that I thought I would be any better at, so it was in a black mood that I first sat down again at the microscope.

I started looking at some of the bigger cells on the outside of the embryos, and began to assemble more fragments of the story. Gradually I developed a schedule of planned sessions, working around the embryo. I worked out that if I was to finish the job, it would take a year and a half of looking down a microscope every day, twice a day, for four hours each time. It seemed crazy, and I consulted John White about whether it was worth doing. 'Am I really going to do this?' I asked him—and John said, 'Yes, sure.' He and Eileen, after all, were spending year after year poring over electron micrographs in order to produce the complete wiring diagram of the worm's nervous system.

So, to the surprise of some of my colleagues, I shut myself away for a year and a half and devoted myself to the embryonic lineage. Whereas the laser ablation and lineage mutant experiments had been quite social activities—Judith remembers that we talked for up to five hours a day during her first year as a post-doc in 1978–9 – working on the embryonic lineage was an entirely solitary endeavor. I had a little cubby-hole off the main lab to do the microscope work, and I spent almost all my time in there. Most days would include two sessions of four hours or so watching cell divisions. With the postembryonic lineage you could afford to leave the microscope for ten minutes or so in the course of watching a cell, but not with the embryo—things happened too fast. You had to concentrate totally. One difficulty was making sure that I didn't become muddled about which cell was which: they all look alike. So I made a classic cross-hair from spider's gossamer and used it to pinpoint a cell in the area I wanted to watch. Then viewing was much more relaxed.

After a year and a half it was done. Many people wouldn't have seen the point; many still don't. In molecular biology, tasks of this kind are called 'Hershey heaven', after a remark of Alfred Hershey, one of the key members of the group who laid the foundations for much of modern genetics in the 1940s and 1950s through their studies of tiny viruses called bacteriophage: you come in every day, you do the same experiment, and it always works. It's rather like doing a huge jigsaw puzzle—it may be difficult, but you finish it off, partly because it's the best possible excuse for saying, 'Don't bother me, I'm busy,' but also because you think it's important. And it goes better and better because you're training yourself to do that particular task exceptionally well. That it's important is crucial, though: there's no point in continuing with something that's turning out to be valueless. With very big projects, if you're collecting very high-quality data, which you know contains a lot of information even though you can't get it all out at the time, then it's worth pursuing to the end.

That was true of the worm lineage, just as it later proved true of the genome sequence. The lineage is now a resource that people refer to all the time. Coupled with the anatomy, it makes it possible to identify the sites where genes act. The lineage, to quote Bob Horvitz, 'has given single-cell resolution to worm biology.' And it's not relevant only to worms. We discovered, for example, that certain cells consistently die during development. Bob Horvitz and his colleagues went on to unravel the genetic control of this form of cell death, which turns out to have parallels in many other species, including humans. Activating the cell death mechanism could be a route to treating cancers, while suppressing it might help in the treatment of degenerative diseases.

Some time after I'd finished with the embryo, John White invented a recording device, using an optical disk, that really did allow the embryonic lineage to be reconstructed without direct viewing. And others are now trying to develop a fully automatic method using fluorescent markers. Sometimes people say: 'Aren't you sorry you wasted your time?' But of course I didn't waste my time; you have to make a start somehow, and then eventually the first approach is displaced by better ones.

To be in Sydney's division at the LMB in the 1970s was to be in at the birth of an international community of worm biologists. The great majority of today's worm researchers either came to work with Sydney or are scientific descendants of those who did. Bob Horvitz, now at MIT, and Bob Waterston, now at Washington University in St. Louis and my closest collaborator on both the worm and the human genomes, were among the early ones. Judith Kimble left to start her own lab at the University of Wisconsin–Madison. John White stayed at the LMB until 1993; now he, like Judith, has a lab in Madison. Donna Albertson came as a visitor, married John, and stayed on. Jonathan Hodgkin was the third of the three Jo(h)ns, and we four were more or less permanent fixtures in the lab while others came and went. One of Sydney's first Ph.D. students, Jonathan stayed at the LMB until very recently, when he moved to Oxford as professor of genetics. There are dozens of others, and the worm community as a whole now has thousands of members. But it retains the extraordinary community spirit that developed at the LMB.

Sydney's personal influence was undoubtedly hugely important, but the nature of it is hard to pin down. His notion of supervision was to throw out the odd idea and then go away and leave you to it. There would be no weekly lab sessions to check on progress; but he might turn up and quiz you at odd hours. He would talk about nothing in particular for hours in the coffee room or the corridor, but if you needed to talk seriously to him he would suddenly become elusive. I have vivid memories of trying to have conversations with him through a pair of closing lift doors as he left at the end of the day. He was never in a hurry to publish—it was five years after my arrival at the LMB that we finally published our first paper together, on worm DNA. He is not someone who suffers fools gladly, and he was a master of the cutting put-down. I treasure the reply he gave when I told him my anxieties about the tenure issue, saying, 'I don't want to carry the can for all of this.' Said he: 'Don't worry, I'll carry your little can for you.' His is a complex and powerful personality, and he dealt ruthlessly with anyone he saw as a potential competitor. But he single-handedly started *C. elegans* research, and we all learned a lot from him.

People congregated in the coffee room at certain times of day. After lunch, Sydney would hold court, talking for an hour on any topic with the young researchers gathered at his feet. Coffee-time in the morning or tea-time in the afternoon were opportunities to talk about science with anyone who was passing through. After the first year or two I tended not to go to these gatherings so often, but Friday evening at the hospital bar, the Frank Lee, became a regular drinking session. One evening in the summer we used to go punting up the river and have supper at the Green Man at Grantchester. We'd come back down well tanked up, flopping in and out of the river, letting the punts float down more or less on their own, with candles on the front. The Cambridge worm group does much the same to this day. Another fixture was the Guy Fawkes night celebrations, with a bonfire and fireworks, held round at our house every year for years and years. Lots of worm people remember those evenings with tremendous pleasure. I've no doubt it's one of the activities that built a community. No one thing was essential, but people used to talk until quite recently about coming back to visit 'Mecca'—the worm group at LMB.

2 ON THE MAP

LOOKING BACK, THERE ARE A FEW OCCASIONS THAT YOU THINK OF as pivotal events that changed your life. For me, agreeing to sequence the worm genome was one of those, but in many ways the critical time was earlier. I was at a conference on developmental biology in the United States, listening to Matt Scott, a researcher from the University of Indiana who worked on the fruit fly *Drosophila*. Matt's group had been working for some time to map the *antennapedia* region of *Drosophila*—a region of the genome where mutations cause the fly to grow a leg on its head instead of an antenna. Understanding mutations such as this is fundamental to understanding the normal process of development that puts legs and antennae in all the right places. They'd done it by chromosome walking: working sequentially along part of the chromosome. And just listening to him describe his beautiful work, I could see with stark clarity that if you could cover the genome in parallel instead of serially, then for only a few times more effort you could map the whole thing.

Mapping is a step towards understanding the flow of information from the coded instructions in our genes to the molecular interactions that go on within and between our cells. It is these inter-actions that underlie all the functions of the living body: burning food, fighting infections, healing wounds, even running the 100 meters or composing a symphony. At the LMB we were trying to put together a picture of how a very simple animal, the nematode worm, is programmed to grow from egg to adult and to carry out its essential tasks of moving, feeding and reproducing. Those who worked on *Drosophila* were trying to do the same with a much more complex animal, one that has eyes and legs like us—and wings, too.

Achieving a fuller understanding of life in all its richness and complexity is immensely interesting for its own sake, just as it is worth struggling to understand the origins of the universe. But molecular biology—even of worms and flies—also offers huge potential spin-offs for human health. Right up to the end of the twentieth century much of modern medicine was based on a rather hit-and-miss approach to finding what works. Antibiotics are a classic example: Alexander Fleming's accidental observation that a mould could kill bacteria led Howard Florey and Ernst Chain to extract and purify the active agent in the mould, penicillin, and use it to cure patients with infections. But thousands of lives had been saved before anyone understood *how* the drug killed bacteria. For the past couple of decades the aim of the pharmaceutical industry has been to replace this serendipity with a more rational approach in which treatments are based on what we know about how living things are put together. The techniques of molecular biology potentially make it possible to understand the difference between sickness and health right down at the level of one molecule inter-acting with another. And many of these molecular interactions are common to animals as simple as a worm and as complex as a human.

A crucial first step is to find the genes that control these inter-actions. Each gene is typically thousands of bases long, and its sequence of As, Ts, Cs and Gs usually encodes a protein. The code is a set of three-letter words—TTT, CAG and so on—each of which corresponds to one of the twenty amino acids that are the building blocks of proteins. It's the proteins—there are hundreds of thousands of them, but some better-known examples are insulin, pepsin, hemoglobin and keratin—that actually carry out the business of building and maintaining a living body. Each human gene has its place on one of the twenty-four chromosomes (numbered 1–22, plus the X and Y sex chromosomes), which together constitute the whole human genome. Some genes are on one strand of DNA and some on the other, read in opposite directions. Finding them is not easy. We now know that only about 1.5 percent of the total DNA actually codes for proteins. The rest is often pejoratively termed 'junk' DNA, though it is more accurate to call it non-coding DNA: much of it may well be junk, but scattered in it are all manner of control sequences to which signalling proteins can bind, causing genes to be turned on or off as required and defining the stop and start positions. These controls are vital, because nearly all of our 100 million million cells contain exactly the same DNA, yet each has a specialized job to do. When a gene is turned on, the appropriate stretch of sequence is transcribed into a single strand of another sort of nucleic acid called RNA. The RNA code is very similar to that of DNA, with minor chemical differences. The RNA transcript then moves out of the nucleus, and its nucleic acid code is translated into the amino acid chain of a protein.

The junk itself is a collection of fossils from our evolutionary history, which makes it interesting in the same way that a midden is interesting to archeologists. To make life even more complicated, the coding part of most genes is split into numerous segments called exons, separated by much longer non-coding sections known as introns. Some small genes on one strand can actually sit inside the introns of other, larger genes on the other. You can begin to see why finding genes is a far from simple matter.

Now, in order to find things it is useful to have a map. A map allows you to home in on the place you want and avoid going round in circles. In the search for genes, researchers have used two kinds of map: genetic maps and physical maps. Genetic maps have been made for almost a century; the cartographers were the geneticists who bred generations of mutant fruit flies or other species in order to track the mutations through the generations and work out how closely they were linked (just as Sydney did with the worm mutants). None of this work involved handling the DNA itself; it was almost 1950 before people generally accepted that DNA was the stuff of inheritance, and until the early 1970s the tools were simply not available to begin to analyze it in detail. So a genetic map is an abstract entity that tells you the relative positions of genes on chromosomes. When geneticists say they have 'found the gene' for a particular trait or disease, they usually mean they have placed it on a genetic map.

Finding the actual piece of DNA that constitutes a gene requires a physical map. Physical maps are much newer than genetic maps. The essential tools to make them were restriction enzymes, used to cut the DNA into fragments a few tens or hundreds of thousands of bases long, and cloning techniques, enabling each fragment to be inserted into a bacterium and multiplied as the bacterium grew into a colony. A physical map is a collection of cloned DNA fragments that have been arranged in the right order along the chromosome by looking for overlaps between them.

Mapping becomes really powerful when you line physical maps up with genetic maps. Then, once some genes are located on partic- ular fragments, you can make a good guess about where intervening genes are located, and so the whole system becomes more powerful as one proceeds. The combination of the two is a genomic map. With a genomic map, you don't just know the location of a gene as an abstract point on a diagram; you know that a particular colony of bacteria in your freezer contains that gene within the clone it har- bors, and so the sequence of the gene itself is within your reach.

By the 1980s genetic maps were good enough for human

geneticists to go fishing for genes among the families of people with inherited diseases such as Huntington's disease or cystic fibrosis. The maps were staked out with markers, short sequences of DNA that come in two or three different varieties in the population. If it turned out that everyone in a family who had a particular variety of marker also had the disease, then there was a high probability that the gene responsible was on the same chromosome as that marker and close to it. Having narrowed down the location of the gene in this way, geneticists then switched to 'walking' to try to clone the gene with a view to reading its sequence and understanding how it works.

In order to walk part of the genome you lay out clones covering the whole genome (or sometimes a chromosome) on a membrane, and 'probe' them with a radioactively labelled sequence from one of the marker clones. You pick those clones that stick to the probe, and analyze them to see which extends furthest in the direction you want to go. At first, even knowing which way to go is unclear, so you have to go both ways. You then choose a piece of sequence from the far end of the new clone to use as a probe, put a radioactive label on it, and repeat. As you proceed, you find more markers, get the direction sorted out, and start to form a physical map. With walking, mapping is a serial process—you can't map the next clone until you've done the one before it. It was slow, but it was this method that in the late 1980s and early 1990s tracked down a number of genes altered in disease, for example those for cystic fibrosis, muscular dystrophy and Huntington's disease.

Matt Scott was doing the same kind of thing to try to track down the faulty gene that makes fruit flies grow legs on their heads. What I thought, listening to Matt's talk at that conference, was that you could do the whole process in parallel. Instead of picking the clones one by one, you could take all of them and characterize the whole lot in some way that would allow you to detect overlaps in a computer. When I first had this thought I was unsure how to begin; I'd been looking at cells down my microscope for the previous eight years and didn't know anything about molecular biology. So I went to talk to Sydney and to one of his colleagues, Jon Karn, for advice about exactly what I would need to do to characterize the clones. Jon, who was working on the molecular genetics of worm muscle, suggested a method which we came to call fingerprinting. (Our method was not the same as the DNA fingerprinting developed two years later by Alec Jeffreys, and now used in forensic testing and paternity testing, but in both cases we were after a way of uniquely identifying a sample of DNA.) I was vaguely aware that I could fingerprint a clone by treating it with an enzyme that cut it at specific sequences and then sorting the resulting pieces by size. A standard lab technique for sorting mixtures of biological molecules is gel electrophoresis. Various jelly-like substances can act as molecular sieves. If you place the gel in an electric field the molecules move through it, sorting themselves out from smallest to largest (smaller pieces move faster than larger), giving a unique pattern of bands on the gel.

Jon's method was slightly more sophisticated and gave better resolution, because it generated smaller fragments that could be run through a different sort of gel. It involved two different enzymes and a radioactive label that marked the ends of the pieces of DNA after the first cut; the second cut broke the DNA into smaller fragments ready to run through the gel. Exposing the gel to photographic film would give me a pattern of dark bands like a bar code, showing the positions of the radioactively labelled fragments. Each clone would have a unique bar code, because the enzymes would cut it at specific sequences that they recognized. The bands were not in order, but that didn't matter. The idea was simply to look for partial matches between the bar codes of different clones, which would indicate an overlap. Once we had overlapping clones for the whole genome, we would have a complete map.

I saw that by digitizing the fingerprint information it would be possible to hunt for overlaps automatically in a computer. Instead of crawling along the genome clone by clone, you could map the whole thing in one go. That was the point at which I discovered for myself the power of genomics. I had just finished the embryonic lineage, and was looking for something else to do. And at that point I was driven by an obsession that to map the worm genome was the right thing to choose. Just as the lineage had been a resource for developmental biologists, a genome map would be a godsend to worm biologists who were looking for genes. I wasn't interested in looking for genes myself. Nor was I particularly keen to specialize further in the developmental biology of the worm by looking for lineage mutants. I was just captivated by the idea that here was another opportunity to map a big chunk of the biological landscape.

Simply gathering data without having any specific question in mind is an approach to science that many people are doubtful about. Modern science is supposed to be mostly 'hypothesis driven'—you have a hunch about how the world works, and do experiments that ask if your hunch is right. If it is, you can make predictions about how the world might work in other, similar situations. My first studies of the worm lineage didn't require me to ask a question (other than 'What happens next?'). They were pure observation, gathering data for the sake of seeing the whole picture. Making a worm map would be the same. This is sometimes called 'ignorance-driven' or, more grandly, 'Baconian science.' The seventeenth-century philosopher Francis Bacon suggested a system for understanding the world that began with the accumulation of sets of facts, based on observation. Naturalists who collect and classify living species or astronomers who map the stars in the sky are examples of Baconian scientists. This kind of project suits me—it's never bothered me that it doesn't involve bold theories or sudden leaps of understanding, or indeed that it doesn't usually attract the same level of recognition as they do.

First I had to find somewhere to work. I'd already relinquished my cell lineage space, but Sydney, who by this time had succeeded Max Perutz as director of the LMB, had a vacant room in the ominous-sounding extension of the lab called Block 7. It was Room 6024, and that number is etched on my mind as the place where in 1983 worm genomics began.

Jon Karn told me that he'd heard via Bob Waterston that Maynard Olson was thinking along similar lines, in order to map the yeast genome. Maynard was working in the genetics department at Washington University in St. Louis. It was a year or so before I met him. An austere figure with heavy-framed spectacles, Maynard came from a very different scientific tradition from me, in which new methods were thoroughly worked through theoretically in advance before you ventured to start an experiment. I was always much more keen to get into the lab and try things out. But despite our differences it was straight away good to talk to a fellow believer, and Maynard also appreciated the benefits of a 'mutual support group' of genome enthusiasts. We kept in regular touch over the next few years; later, as the Human Genome Project gathered pace, Maynard's capacity to analyze the possible outcomes of different courses of action made him a respected voice in the occasionally fractious debates about strategy that blew up from time to time.

Sydney Brenner and Jon Karn were supportive, of course, but most people were skeptical of this enterprise that had nothing directly to do with biological problems. Maynard told me that most people he talked to went round on a circular path of criticism. 'First they say it can't be done. Then when I've dealt with their objections, they say "Well, but there's no point to it." So I explain exactly why it's valuable, and they say, "But it won't work, because..."' I was having the same experience. And indeed it wasn't very successful at first. I remember in particular a *Drosophila* post-doc whom I knew coming in and saying, 'What on earth are you doing this for?' She was really quite angry. She looked at my messy filters; it obviously wasn't working very well. She said, 'You've got the embryonic lineage of the worm in your head. You ought to be sitting at the microscope picking worm mutants.' I felt like a little boy in short trousers, but I said, 'I'm sure you're right, but I want to do this—I think this is important.' Another *Drosophila* researcher said dismissively, 'It'll all be over in five years—we'll have solved developmental biology in the fly.' It was true that in a spectacular development in fly genetics, Christiane Nüsslein-Volhard and her colleagues in Tübingen, Germany had described a whole lot of *Drosophila* mutants that disrupted the normal development of the animal's body plan. But I said, 'I don't think so; it's more complicated than that. We're going to need all the genes.' And that was the reason for doing the map. I felt that we had to go for more than could be done by looking for mutants. With mutants alone, one would not be able to see everything; if a mutation was lethal, for example, or if more than one gene carried out the same function, the link from gene to function would be hard to analyze or undetectable. But even just speeding up the isolation of known genes was going to be valuable in itself.

The phrase that was in my mind was: 'We're going to clone all the *uncs*'—the uncoordinated mutants of the worm, of which Sydney and his colleagues had identified more than a hundred. The interesting thing about the *uncs* is that they can have all sorts of things wrong with them—it could be muscle, or nerve, or something else—but among them are going to be interesting, complex things about the worm that are the sorts of things that we might be interested in finding in higher animals. But people were cloning very slowly—it took a lab years to clone a gene. John White remembers how the excitement of discovery that pervaded the lab began to deteriorate as everyone began to try to clone genes.

> Lab meetings became nothing but progress in mapping. It was absolutely mind-numbing. The field became less interesting because all people cared about was cloning and they forgot what the biology was behind it.

John and I had occasionally talked about this problem during our Friday evening sessions in the pub. And before long I realized that if we could make the map, we could then give the geneticists all the *uncs*—they'd all be cloned, just like that. And that's how it turned out. Not quite as easily as I'd hoped, but a few years later we'd done it. All the *uncs* were cloned for anybody who wanted them, though it was easier to pull out the ones where there was a higher density of genetic markers that could be used to align the maps.

The mapping project did not really take off until Alan came. Alan Coulson had been a research officer working with Fred Sanger in the LMB on developing first RNA and then DNA sequencing techniques. Having been Fred's assistant since leaving college in 1967, he had worked indefatigably in the late 1970s on developing and implementing Fred's dideoxy, or chain termination method of reading DNA sequences.

Fred's method is essentially the same one that all sequencers use today, although improved chemistry, miniaturization and automation have speeded it up immeasurably. He used a piece of single-stranded DNA a few hundred bases long as a template. The template was used to produce many complementary copies, by means of one of the polymerase enzymes that copy DNA in real life: the enzyme pairs Cs with Gs and As with Ts, all the way along the strand. Each copy began from the same point, defined by a short 'primer' fragment of DNA that was bound to that spot and nowhere else. The enzymes gradually added bases to the primer fragment from a pool of all four normal bases spiked with an altered, 'dideoxy' version of one of them. For each template there were four different reaction tubes, each with just one of the four dideoxy bases. When the enzyme constructing the copy randomly added a dideoxy base instead of a normal one, it stopped the chain. The four mixtures were run side by side through a gel to separate them by size, giving a set of 'ladders' with unevenly spaced rungs showing the positions of all the dideoxy bases relative to one another. Taking the four

47

ladders together, there was just one rung at each successive level, and by noting which tube each rung had appeared from Fred and his colleagues could tell which base it represented and so could 'read' the sequence.

Using this technique they successfully carried out several of the first ever whole-genome sequencing projects, each more ambitious than the last. They would begin by 'shotgunning' the genomic DNA into random fragments, and carry out a set of sequencing reactions for each fragment, so generating from each a sequence read. They had no advance knowledge of where each read would fall, and simply continued randomly until they had 'coverage' of tenfold or so—that is to say, there were 10 times as many reads as would be needed to cover the genome if fitted end-to-end precisely. Then, by looking for matches between the reads, they could 'assemble' them into a composite, which was the sequence of the genome. Additional work would be needed to fill gaps and resolve poorly read sections.

To begin with, they assembled the reads with pencil and paper in their notebooks. But it soon became apparent that this was a job for computers, and so Rodger Staden, an assistant from the LMB's structural studies division, was 'borrowed' to write some of the first programs designed to look for matches in the sequence that would reveal overlaps between one fragment and another. The first organism to be sequenced was bacteriophage phiX174, a tiny virus that infects bacteria, with a genome of about 5,000 bases. They then moved on to the DNA in the human mitochondrion (the energy-generating part of the cell) with about 17,000, and bacteriophage lambda with nearly 50,000—each time increasing the size by about a factor of three. All these were successfully completed using the time-consuming manual methods of the day. Alan realized that genomes much bigger than these would be too unwieldy to sequence straight off with existing technology. Fred's previous research officer Bart Barrell, who had joined Fred from school in 1963 and was now working independently at the LMB, was using the method for viral genomes with up to a quarter of a million base pairs, but that seemed to be the limit for the moment.

When Fred Sanger retired in 1983, Sydney Brenner suggested that Alan Coulson might like to come and work with me. Alan was interested to hear that I was planning to map the worm genome, which at 100 megabases seemed at that time impossibly large for sequencing. Mapping seemed like a more realistic approach to understanding larger genomes, 'a logical extension of what I had been doing with Fred', says Alan. Fred had been a hero of mine since I read about his work on protein sequencing as a schoolboy in the 1950s, and it was a strange quirk of history that his assistant was about to become my colleague. Although we had worked in the same building for fourteen years, I'm not sure we had ever spoken before. Alan remembers that each division had its own characteristic style:

> I saw the crystallographic people on the ground floor as sports jackets and brogues. The middle floor, cell biology [where John was] was much more sandals and beards. And Fred's group, protein and nucleic acid chemistry, we were on the top floor—and we were just normal!

So Alan—tall, bearded, soft-spoken and self-effacing—came and met me, and we looked each other up and down. We started to chat about what he might do, and then I suggested we continue the conversation in the pub. For Alan, it was an early indication that working with me was going to be different from working for Fred, who was rather more formal in his approach. I'm not a good manager, but I do like partnerships; I assumed that we'd just share things and do everything together. And so it turned out. We worked so well together that we tended to be regarded as a unit, John'n'Alan. Once a technician in Bob Waterston's lab in St. Louis was vainly looking for our e-mail address, and it turned out that he was trying to find the address of someone called John N. Allen.

We settled into a routine. I prepared the clones that we would need as a source for the map. In order to make clones, you need a vector—a piece of DNA that can be opened up to accept an insert of worm DNA, and which will in turn carry the worm DNA into bacteria where it will be replicated as the bacteria multiply. For our map we used vectors called cosmids which could accept 40 kilobases of DNA. Then Alan fingerprinted the clones and ran the gels. Our strategy depended on digitizing the images so that they could be analyzed by computer, and that was my main task. I began by using a manual device developed by Rodger Staden. It involved accurately hitting each band with a stylus, with the computer automatically recording the position. Rodger also wrote us a simple program that would search for matches. Once we had the data, Alan would assemble the clones into 'contigs'—Rodger's word for regions of the genome covered by overlapping, or contiguous, clones. The idea is to end up with a small number of large contigs—in a perfect world it would be one per chromosome—although of course at the beginning of a project you have a very large number of small contigs. It's very satisfying when the number of contigs starts to fall as they link up with one another.

We scanned hundreds of clones this way, but the method was clearly going to be too limited for the whole job. Scanning with the stylus was tedious and depended on the accuracy of the operator, and the output was just a list of matches with no means of manipulating those further. Alan recorded the contigs as pencil lines in his notebooks which he rubbed out and redrew as we found more matches. We needed an electronic alternative, and as by this time Rodger Staden was busy with his thesis, I decided to make it myself. I learned some Fortran (at that time the standard scientific programming language), talked to Rodger, and, just as Sydney had, found myself completely hooked on programming. What I built was a graphics-based program called contig9 that represented the clones by lines on the computer screen. All Alan had to do was position two contigs, press a key and they would join. Later I made the matching part more automatic. The program gradually grew into a large piece of code, developing from what Alan and I found we needed as we went along.

We also needed a way of automatically reading in the data from the films. This was more difficult. There were no commercial scanners available then, so the LMB workshop built one. I took on the job of putting together the associated software. It was important to have an automatic program that could read the bands in the sample lanes and decide how they lined up with the marker lanes— which had many bands, rather like a ruler—that we ran alongside as a point of reference. It wasn't easy because each gel is slightly different; they tend to distort slightly, so that the positions are never exact.

At the time I was wrestling with this problem, John White was supervising a very bright graduate student, Richard Durbin. Richard is now the deputy director of the Sanger Centre and has played a key role in developing the software without which large-scale genome projects would have been impossible. He is immensely thoughtful and weighs his words carefully before committing himself. (One day—this was years later, when we were at the Sanger Centre—I was chairing a management meeting, and, having had my say on one of the agenda items, popped out to the copier to make some duplicates for the next one. When I came back the room was silent, so I assumed they'd finished, dished out the copies, and moved on. After a moment, the others gently interrupted me to say that Richard was in the middle of a sentence. He gets away with it, because everyone knows it's worth waiting to hear what he has to say.)

When I started work on the worm map Richard was doing a Ph.D. on the nervous system of the worm, but he had a mathematics degree and had worked for a year before that in the computing industry. While doing his Ph.D. he had written the software for the confocal microscope which John White had designed, a beautiful piece of work. I took him my problem with the lining up of the bands, and he solved it in no time by contributing a dynamic programming algorithm that considers all possible matches and picks the most likely one. I built in nice, easy-to-use editing arrangements so that I could look at the scan, check immediately whether the algorithm had picked the bands correctly, and, if not, make a small edit. I learned how powerful it is to have an interface where the computer does all the clerical work for you and presents an easy-to-edit image. There was nothing theoretical or fancy in the software, but it was just what we needed to get the job done.

It was very important to us from the start that the whole community of worm researchers should be involved in what we were doing. Without this small but enthusiastic group (at that time there were probably not more than a couple of hundred worm people in the world—maybe a tenth of the number working on the fly), the information in the map would be meaningless. They would provide the genetic markers from known locations, so that the physical map could be aligned with the genetic map. Once enough of these markers appeared on the map, people would know where to look for particular genes that they were interested in—and we could supply them with the clones that they needed to make the search.

One of the leading members of the worm community was Bob Waterston. After training in medicine at the University of Chicago he had come to the LMB as one of Sydney's post-docs in 1972. He had worked on *uncs* that had mutations affecting their muscles, and had gone back to set up his own lab at Washington University in St. Louis, using the worm to study the genetics of muscle. 'It was clear that the way we were going to learn about muscle was by cloning genes,' he says. 'But it was really slow going—we needed a new means to do it.' In the laboratory next door to Bob, Maynard Olson was making progress with his yeast map. Bob and

Maynard were good friends, so Bob was aware early on of the idea of taking a systematic approach to understanding an organism's biology, and he was excited when he heard what we were up to. When he had the opportunity to take a sabbatical, he wrote to ask Sydney if he could come back for a year. He arrived to find that there was nowhere to sit in Sydney's main worm group, where he had planned to work on embryonic muscle mutants. And so he, too, came to Room 6024.

Bob is lean and muscular, with a moustache and a fringe of wavy red-gold hair around a bald crown. His most outstanding character-istics are his unfailing amiability and his total honesty; he's also clever and wise. This was the second time he had come back on sabbatical, but during his previous LMB visit I was so absorbed in the lineage that I didn't socialize much, hardly ever going to communal coffee and tea sessions, so I hadn't seen much of him. And then in 1985 he came and just sat there with Alan and me, and we talked about the genome.

We had a problem at the time Bob came. For reasons we didn't understand, some parts of the worm sequence just refused to clone as cosmids, so that there were gaps in our map—about 700 of them. Worm biologists were already finding the longer contigs useful, but with so many gaps the smaller pieces were unattached. We needed to join the map up fully. Bob experimented with various methods of linking the clones while he was with us, but none of them worked. Towards the end of his year in Cambridge he went back to St. Louis to see how his lab was doing. While he was there, he dropped into Maynard Olson's lab and discovered that one of Maynard's post-docs, David Burke, had found a way of cloning much longer pieces of DNA in yeast cells. He called these clones yeast artificial chromo-somes, or YACs. Bob thought that YAC clones might be able to bridge the gaps in our cosmid map. Not only were they longer than cosmids, but, more importantly, they grew in a different host: yeast keeps its genetic material coated with protein and contained in a distinct nucleus, like the worm and all higher organisms but unlike bacteria. We thought that the problems with cosmids might be that certain worm sequences were rejected by bacteria, and that these sequences might survive in yeast. Almost at once Bob offered to make a YAC library of the worm genome and to become a partner in our mapping project. 'By that time I had become taken with the idea of the power of the map,' he says, 'so I was happy to participate.' So began another rewarding and fruitful research partnership that endures to this day. Not that we had any idea at the time where it would lead: Bob was just offering to join in on the mapping problem. He made the YACs and sent them to us, and we sent cosmids to him, and we each did experiments looking for matches between the clones to find out which YACs could bridge the gaps in our cosmid map.

Just then Yuji Kohara came from Nagoya University for a sabbatical year to learn about the worm. He had just completed a wonderful map of *E. coli*, using a different method of fingerprinting that he had devised. He had really come to start working on gene expression, but for a few months he willingly joined in, and in that time we aligned all Bob's YACs with the cosmid contigs. We continued steadily making joins, and soon most of the map was in large contigs. Our suspicion that bacteria didn't like parts of worm DNA turned out to be correct.

At this point the genome map became truly useful—and the community of worm biologists came into their own. They used the map to find the genes not just as abstract locations but as physical pieces of DNA. With these in hand they could carry out recombinant DNA experiments to find out how the genes worked, study the expression of the genes in different tissues, make antibodies to the gene products—all the techniques of modern molecular biology. The genes also helped us by providing new landmarks on the map: it was a virtuous circle. And we in the genome labs had determined quite formally among ourselves that we would not use the mapping information for our competitive advantage, in terms of searching for genes. We realized that if we waited until the map was complete before we published it, we would be sitting on a lot of information that was of value to the community. So almost from the start we began to make the mapping data available electronically over the predecessor of the internet. Every so often I would put all the latest map data onto a computer tape and post it to Bob. He would load it on to the computer at Washington University, and then set it up so that people could access it from their local computers. We made the data available to researchers in Europe in the same way. The reason for storing the information in more than one place was that communications were so slow that it made a real difference how far away the database was. This was long before the World Wide Web; we were using bitnet, which could take only small packages, and sometimes, in extreme cases, they took a month to arrive. So I developed a system of incremental updating, to avoid having to send the whole thing on tape every time.

The map was constantly on display. We had regular updates in the *Worm Breeders' Gazette*, the informal newsletter of the worm community; we showed it at conferences; and anyone could request clones at any time, free, immediately, whatever they wanted, so that they could look for genes. There was too much work for Alan and me, and we recruited Ratna Shownkeen—slender and gracious, she is now a project leader in the Sanger Centre. The traffic in worm clones continues unabated to this day. There is no doubt that the science progressed faster because of this two-way exchange of information than it would have done if we had tried to keep the map to ourselves. And the way we handled the worm map set a precedent for our handling of the worm sequence data when it began to flow, and, ultimately, of the human sequence data.

For the whole family, this was a time of moving on. Our children were growing up and about to leave home. Ingrid went to read biochemistry at Leeds, while Adrian in his turn went to study mathematics at Warwick. And for me, the map meant a different way of working. Previously I had always avoided responsibility for anything other than my own work. But the map was not something I could do on my own; it involved commitment to others, such as Alan Coulson and Bob Waterston, to keep the show on the road. At the same time some changes occurred in 1986, the year we published our first paper on the worm map, that propelled me to greater independence. First, Sydney successfully proposed me for election to the Royal Society, the U.K.'s national scientific academy. Although I always hoped that the cell lineage work would be useful, I didn't expect that it would be valued to such an extent by the wider scientific community. Second, Sydney himself retired as director of the LMB and started a new MRC Molecular Genetics Unit where he would work on human DNA (and later with the DNA of the puffer fish *Fugu*, which has a remarkably compact genome with very little 'junk'). This meant a moving apart, because my place was in the LMB and with the worm—and a recognition that from now on I would have to carry my own can.

We embarked on the worm map because it would be a useful tool in itself—there was no need to justify it as a stepping stone to the next step, which would be to read the complete worm sequence. But it certainly wouldn't be true to say that we never thought about sequencing. Bob Horvitz remembers a particularly drunken evening at a Cold Spring Harbor worm meeting in the mid-1980s, when we had embarked on mapping the worm.

> There were four of us who decided to sit in Blackford Hall [the building that houses the Cold Spring Harbor dining and reading rooms] and think about whether it was realistic to determine the sequence of *C. elegans*. And it was John Sulston, and me, and Gary Ruvkun [a colleague of Bob's], and Winship Herr, who was at Cold

Spring Harbor [he is now the deputy director] and supplied the beer. The four of us sat there basically talking through whether, given the technology of the day, it was possible for a small number of people, i.e. John, to determine the sequence of the animal. And we made assumptions, we did calculations, and at the end of the discussion we decided yes, it was feasible. That was unthinkable in the general community. The numbers were just too big, I mean 10^8 base pairs! It might have taken a few years, but we agreed it was not impossible and that John should do it.

Bob claims that the following morning I showed no recollection of this beer-fuelled discussion, but I suppose something must have lodged in the back of my mind. Maynard Olson also remembers similar discussion over dinner when he was on a visit to Cambridge during Bob Waterston's sabbatical in 1985–6.

We talked about genomes and where all this was headed. What stuck with me was that this was the first time John expressed a strong desire to move on to sequencing, not just in the worm but generally. He had a better feeling than I did that it might be feasible to sequence genomes on some kind of timescale that was relevant to our discussion.

From the beginning of the 1980s the idea of moving from mapping to sequencing genomes was being aired in relation not only to small organisms but to the human. Sequencing would provide the ultimate in biological information. Already the sequencing of individual genes was revealing an intriguing picture: there was a very high level of similarity between worm, fly and human genes that did the same jobs. Throughout evolution, mechanisms that work seem to have been conserved almost unchanged. Genome sequencing would provide the means to extend these comparisons, using the simpler organisms as a window into what goes on in human cells.

But if the worm seemed beyond reach at 100 megabases, how could anyone do the human at 3,000 megabases? One of the first to dare to think on this scale was Robert Sinsheimer, who was chancellor of the University of California at Santa Cruz. Sinsheimer was a molecular biologist who had isolated, purified and mapped the DNA of the phage phiX174, the very first organism to have its whole genome sequenced (by Fred Sanger and his colleagues). By this time more of an administrator than a bench scientist, he had been involved in the efforts of the University of California's astronomers to raise the funds for a huge new telescope. It was ultimately the California Institute of Technology that secured the crucial donation, from the Keck Foundation, to build the Keck Telescope on Hawaii at a cost of more than $70 million. As a result, the University of California lost a $36 million donation from the Hoffmann Foundation, which had earlier hoped to name the telescope. In 1984 Sinsheimer began to wonder why such large sums should not be raised for biology. 'I wondered if there were scientific opportunities in biology that were being overlooked, simply because we were not thinking on an adequate scale,' he said.

He conceived the idea that a sufficiently ambitious biological project might be able to win back the Hoffmann donation for the University of California. He consulted colleagues in the biology faculty about the feasibility of an Institute to Sequence the Human Genome and so put Santa Cruz 'on the map.' Among those he consulted was Bob Edgar, a worm biologist who had worked in Sydney's lab and knew at first hand what Fred Sanger had achieved. A letter went off to Cambridge seeking Fred's opinion. He replied, '[It] will probably need to be done eventually, so why not start now? . . . I think the time is ripe.'

As a first step, Sinsheimer convened a workshop at Santa Cruz in May 1985 to which he invited a mixed group of around a dozen scientists who had some expertise in DNA mapping, automated sequencing or data management. Sydney was on the list but couldn't go, so he sent me and Bart Barrell, head of large-scale sequencing at the LMB. Others included Walter Gilbert of Harvard University, and Leroy Hood from the California Institute of Technology, whose laboratory was making steady progress towards automated DNA sequencing machines that used fluorescent dyes instead of radio-activity to label the fragments. I felt amazed that we were all sitting there discussing making an attack on the human genome. But at the same time I felt confident that we knew what we were doing in mapping the worm, and that if necessary we could scale up the approach to map the human. In retrospect I was probably wrong to think that we could have done it in cosmids—there are extra problems in cloning human DNA that we didn't have to worry about in the worm—but in principle I would have been happy to sign up to mapping the human at that point, because I knew we would have found a way.

Sinsheimer recalled that 'as we analyzed the problems to be solved and the likelihood of progress towards their resolution, the mood of the participants swung from extreme skepticism to confidence in the feasibility of such a program.' Whether it *should* be done was another question: there was a lot of suspicion of the 'big science' approach, and some doubted the value of sequencing the 98 percent or more of the genome that does not code for protein. But overall the con-clusions of the workshop, which Sinsheimer wrote up in a short report, were positive. He thought it made sense systematically to develop genetic and physical maps of the genome, and singled out the worm map as evidence that 'the technique clearly would permit development of a physical map for the human genome within 3–5 years by a reasonably sized group (20 people).' At an estimated rate of 100,000 bases per person per year (and this was itself a very optimistic figure; Bart put the productivity of his own group at about half that, and he was one of the most experienced sequencers in the world), a complete human sequence was not deemed feasible. Instead, the report suggested that the emphasis should be on sequencing 'regions of expected interest', such as genes and genetic markers, until expected improvements in technology for high-throughput sequencing became available.

Robert Sinsheimer, thwarted by the internal politics of the University of California system, never got his genome institute. But he circulated the workshop report to United States funding bodies such as the Howard Hughes Medical Institute, the Department of Energy and the National Institutes of Health, where it added to a groundswell that was beginning to emerge in favor of making a concerted approach to the human genome. One of the leading advocates was Walter Gilbert, who launched what was to become a tradition of hyperbole in the field by calling the total human sequence 'the grail of human genetics ... an incomparable tool for the investigation of every aspect of human function.' Over the next few years a series of meetings ensued. Charles de Lisi, head of the Department of Energy's Office of Health and Environmental Research, convened a workshop at Santa Fe in February 1986, and went on to draw up plans for a genome sequencing initiative that would give new purpose to his department's national laboratories (their interest in genes came out of studies of the effects of radiation). In June the same year a Cold Spring Harbor symposium on 'The Molecular Biology of *Homo sapiens*' found the idea being discussed for the first time in front of a large audience, more than 300 of the world's top human geneticists and molecular biologists. At an informal discussion convened there at the last minute, Walter Gilbert's estimate that the project could cost $3 billion ($1 per base) caused uproar: many of his listeners assumed that funding for biological research would essentially be diverted to this one goal, leaving nothing for the traditional, bottom-up approach to science funding that favored individual innovation.

But although there were reservations within the scientific community, the United States Congress had by now seized on the idea with enthusiasm, so that the momentum for a structured approach to the genome was unstoppable. The National Academy of Sciences (the United States equivalent of the Royal Society), through its National Research Council, set up a panel to examine the whole question of an international genome effort. It was chaired by Bruce Alberts, a molecular biologist from the University of California at San Francisco (now president of the National Academy of Sciences), and its members included Jim Watson, Sydney Brenner, Walter Gilbert and many other luminaries in the field. Meanwhile the National Institutes of Health (NIH) was unhappy about the Department of Energy taking the lead and became increasingly involved in discussions about the funding of a large-scale genome initiative.

As part of this process I was invited to a workshop convened by the United States government's Office of Technology Assessment in the summer of 1987 to talk about the potential costs of the project. I found myself suddenly drawn into a sharp exchange between Jim Watson and Ruth Kirschstein, who at that time was head of one of the NIH's component institutes, the National Institute of General Medical Sciences. This institute had been given responsibility for dispensing grants for genome research, but Jim felt that there needed to be a much more proactive approach if the project was to have any coherence. He specifically argued that one person, and a scientist rather than an administrator, should be put in charge of the program, and unexpectedly turned to me for support. 'Doesn't one person really have to finish up that last 10 percent and live or die for the thing?' he asked. I demurred, worried about giving so much power to one person, but Jim countered, 'Someone has got to do it.' I accused him of wanting to do it himself, which, as the true politician he is, Jim neither confirmed nor denied.

At a meeting on strategy for the genome the following February in Reston, Virginia, Jim again urged that the genome project should be headed by an active scientist. In his absence, several of his fellow participants told the director of the NIH, James Wyngaarden, that

Jim himself was the only credible choice. In May Wyngaarden called Jim to propose that he head an Office of Genome Research; Jim accepted, and the appointment was confirmed that October.

Many people have since wondered why Jim should have wanted to exchange the relative tranquillity of his life as director of the Cold Spring Harbor Laboratory for the bearpit of Washington politics; but, as he himself explained to his Cold Spring Harbor colleagues, 'I would only once have the opportunity to let my scientific life encompass a path from the double helix to the three billion steps of the human genome.' After receiving a favorable report from the Alberts committee, the United States Congress decided to fund genome programs at both the Department of Energy and the National Institutes of Health, but with Jim in charge at the latter there was no question which would be the senior partner. For his first year he had only a planning and advisory role, but in 1989 the Office of Genome Research became the National Center for Human Genome Research, with its own annual budget of almost $60 million. The Human Genome Project officially began in 1990, with a target of a complete human sequence by 2005. Its initial goals were to develop methods and technology through smaller-scale projects, such as the sequencing of simple organisms, before beginning a full-scale assault on the human genome itself. At Jim's urging it also included a program of research into the ethical, legal and social issues raised by genome sequencing.

Although agencies in the United States were furthest ahead in committing serious money to genome research, genome projects were also starting up on a variety of scales in many European countries, as well as in the Soviet Union and Japan. Indeed, a Japanese program had been set up in the early 1980s, recruiting support from a number of private technology companies, to build an automated sequencing facility, and it had been partly a perceived need to keep up with the Japanese that had motivated support for the genome project in the United States. Scientists from the U.K., which had a research record in molecular biology out of all proportion to the country's size or financial resources, were drawn into the earliest discussions about a coordinated human genome project. Bart Barrell and I represented the LMB at the 1985 meeting at Santa Cruz. Walter Bodmer, director of the Imperial Cancer Research Fund laboratories in London and internationally recognized for his work on the genetics of the immune system in relation to cancer, gave the keynote speech at the 1986 Cold Spring Harbor meeting and soon afterwards chaired another critical discussion at the Howard Hughes Medical Institute. Sydney Brenner sat on the National Academy of Sciences panel that built the framework of the Human Genome Project. As well as pushing the project in the United States, Bodmer and Brenner were both active in generating support for genome research at home. Sydney persuaded the MRC to launch a U.K. Human Genome Mapping Project, and was instrumental in the bid made to the Prime Minister Margaret Thatcher, prepared by Professor Keith Peters of her Advisory Committee on Science, for extra government funds.

> I wrote a little note that went in [to the committee], and that became the project. I said we should catalyze and expand the work already being done, especially developing work on computers. We finally got the money; it was not very much, but I thought it was a great accomplishment to get extra money out of Mrs T.

While waiting for the government money to come through, Sydney got started in 1986, using his own funds, working first at the LMB and later at his Molecular Genetics Unit. In February 1989 the U.K.'s Department for Education and Science announced an £11 million grant to the MRC over three years to support the project, with the promise of more to follow. The U.K. project's priorities were very much focused on mapping protein coding regions, especially those that were relevant to disease; its ambitions, like its funding, did not extend to mapping the whole genome and certainly not at this stage to sequencing it.

Walter Bodmer and Sydney Brenner were also instrumental in setting up an international organization with the aim of coordinating genome research. The Human Genome Organization, known as HUGO (Sydney's idea) emerged from a discussion at the 1988 Cold Spring Harbor meeting on genome mapping and sequencing, and was formally founded at a meeting in Switzerland later that year. Bodmer was elected first vice-president and later president. HUGO adopted from the outset a more elitist than inclusive philosophy, with membership by election only. I was 'elected' quite early on—I think Sydney put me in—and I was quite pleased to be associated with the organization; but as I became more involved in it I felt more and more that these people were interested primarily in medical genetics rather than the wider biological importance of genomes. They did not see sequencing the whole genome as the central thing, whereas as far as I was concerned it was going to change everything.

What HUGO did do was to organize regular single-chromosome mapping workshops, at which everyone looking for genes on the same chromosome got together and argued with one another about the positions of markers. It also performed the valuable service of co-ordinating the way new genes were named, and it collected all this information in a Genome Database, originally set up at Johns Hopkins University in Baltimore. Managing the information from genome projects was going to be the key to the success of the enterprise, and the database was in itself a valuable resource. But the genetic mapping data it held could not easily be integrated with the data on gene sequences that was already being collected by another public database, GenBank, funded by the NIH, and its sister databases the European Molecular Biology Laboratory Data Library in Heidelberg and the DNA Data Bank of Japan. The other problem that HUGO never fully resolved was that as an international organization of individual scientists it found it very hard to attract funding. It got started on a grant from the Howard Hughes Medical Institute, and the Wellcome Trust also came in with substantial funds in 1990; but the sums were small in relation to HUGO's ambitions, and other than its coordination of the chromosome workshops and larger biennial meetings, it was never able to establish a position of leadership in the direction of genome research. Inevitably the Human Genome Organization and the Human Genome Project have become intertwined in the public mind; but in practice HUGO played little role in the effort to produce a complete human genome sequence. The push for the genome ultimately came from molecular biology rather than genetics.

The framework of the Human Genome Project emerged under Jim Watson's leadership essentially along the lines recommended by the panel chaired by Bruce Alberts. It would begin with the necessary groundwork and only later move on to large-scale sequencing. The groundwork consisted of making genetic linkage maps, then physical maps that were aligned to the genetic maps; developing new sequencing technologies to increase speed and reduce costs; and testing methods on smaller organisms before moving to the human. Although I had been invited to attend two of the critical meetings that shaped the development of the Human Genome Project, I did not initially think it would have any direct impact on what I was doing. But with hindsight it was obvious that Bob Waterston, Alan Coulson, and I were already doing most of the things that the project was setting out to fund in its early days. We were making a physical map, working with the worm community to align it with genetic maps, and developing technology to allow us to generate data faster and in a form that was easy to use. And in 1989, the year the United States government put its financial weight behind a coherent genome sequencing program, we reached the point in our own project when sequencing suddenly seemed not just possible but the only thing to do. Our map was essentially complete—with the help of the YACs we had got most of the clones into big contigs and were steadily closing the remaining gaps. We presented it at the biennial Cold Spring Harbor *C. elegans* meeting in May. Alan printed it all out on A4 sheets and taped them together, side by side, to form a series of banners, one for each of the six chromosomes. These he stuck one above the other, right across the back of the Bush lecture theatre; and there they stayed throughout the meeting. 'It was impressive,' says Bob Waterston. 'We had a lot of continuity, and we knew how all the pieces lay on the chromosomes. All the other participants at the conference were talking about how they were using data from the map. And that was enormously rewarding.'

Earlier in the year I'd become aware that there were conspiratorial rumblings going on. I was getting messages that Jim Watson was interested in sequencing model organisms before embarking on the human project, and that we had better get our act together if we wanted to participate. Jim had learned of our worm mapping project from Sydney at a Howard Hughes Medical Institute meeting in June 1986, where momentum in support of a human genome sequencing project had begun to build. With the genome project now officially under starter's orders, he was wondering how to promote sequencing. He saw very clearly that the way to convince people of the value of the project, as well as to drive the technology, was to start small. That's the way biology is always done: you don't study humans right off, you begin with something small. By now a series of viruses of increasing size had been sequenced, mainly by Fred Sanger and Bart Barrell, and the time had come to take a hundredfold leap to animals. Jim also knew that it was important to recruit successful labs to the project, both to make rapid progress and to gain the respect of the community. A successful model organism sequencing project would not only act as a trial run for the human, it would perhaps persuade the wider biological community that the genome project was a good idea and not of benefit only to human geneticists.

Our main source on Jim's thinking was Bob Horvitz, my old lineaging partner, who now ran a large worm laboratory at MIT. Bob had known Jim from his time as a Harvard graduate student, and was still in close touch:

> In my conversations with Jim he made it perfectly clear to me that he was thinking of simple organisms, and certainly worms were amongst them. But he wasn't convinced that the worm community had pulled it together enough to launch the operation in the way that would really get the sequence done.

Bob came back to me with the message that there was a list of model organisms that were going to be sequenced, which would include the fruit fly, because it was ahead of the worm in terms of individual genes sequenced. But the worm hadn't quite got on the list—the worm hadn't made it! And that, of course, made us enormously determined to go and get the worm sequenced. Bob Horvitz didn't want to get involved in sequencing himself, but he clearly thought I should. 'We can't muck around—it's a real opportunity and we could lose it,' he told me urgently when we met at another meeting earlier in 1989.

Bob Horvitz now thinks (and Bob Waterston and I agree) that Jim's doubts about the worm were a calculated ploy to sting us into action.

> My impression was that Jim believed the best prospect for serious sequencing was John, and the best way to get John moving was to make him think that if he didn't move the money would go some-place else—like the fly.

If that was Jim's plan, then it worked. We detailed Bob Waterston, who was going to be at a meeting on muscle at Cold Spring Harbor in April, to make sure Jim would be at the worm meeting and would make time to see us. We made sure that he saw our map at the worm meeting. And we booked a private meeting with him in his office for the Saturday evening, just before the close of the conference. We'd had a somewhat hasty discussion beforehand about how to play it, but hadn't begun to think about the details of who would do what. Still, there we were: Bob Waterston, Bob Horvitz, Alan Coulson and me, suddenly doing a deal with Jim Watson about how to start on worm sequencing.

As the meeting was clearly going well, at one point I deployed a tactic that the others had agreed to in advance. 'Look,' I said, 'if you just give us $100 million we'll have it done by 2000.'

Jim barely blinked. He just said, 'That's not the way we do things in this country.'

'Why not?' I wondered privately. But in the end we came to an agreement: between us we would sequence the first 3 million bases (out of 100 million) for $4.5 million during the next three years to show that we were capable of doing it. The work would be divided equally between my lab in Cambridge and Bob Waterston's lab in St. Louis; I would have to seek funding from the Medical Research Council in the U.K., but the National Institutes of Health would fund a third of my costs during this pilot phase to help the British side get started. Jim told us he would be able to justify giving NIH money to a British group for a pilot study on the grounds that it would buy the United States access to the LMB heritage. Always an internationalist, he also believed that the project would be stronger for having an equal partner outside the United States.

At that time 3 megabases was a ridiculous amount of DNA to contemplate sequencing. For comparison, Bart Barrell was just nearing the end of his sequence of the human cytomegalovirus, which at around 240,000 base pairs was the largest genome sequenced to date. It took him five years. But we said, 'All right, we're game; we've got the clones, we'll see what we can do.' None of this was in writing at this time. I didn't go off with a check from

Jim, but I did go off with his verbal promise, and it worked out exactly the way he said. Impressive.

I returned from the United States in tremendous excitement, and came straight from the plane to the LMB to find Aaron Klug, who had replaced Sydney as director. Aaron is a structural biologist, a mild-mannered but decisive man who ran the lab in the same way Max Perutz had—as a chairman rather than a director. He has always been very supportive of me, but when I said, 'We're going to sequence the worm,' his initial response was not encouraging. He said, 'Oh, no!' He knew much better than I did what the task implied—the huge amounts of money, the unremitting work. Then he said, 'All right, if you really want to—but why don't you do the fly? The fly's much more useful.' And the answer was simply that I was not in the fly, I was in the worm. The worm was a much less competitive area; the fly map was a mess, the people who did the YACs were in competition with the people who did the cosmids, everybody was competing with everybody else, it was hopeless. To go into this Colosseum of gladiators was out of the question. The worm had got us where we were—the worm map was the most advanced animal genome map in the world. All I could say to Aaron was, 'Yes, the first animal to be sequenced ought to be the fly, but it's going to be the worm.' There were fly people at the LMB, but they were cell biologists, not mappers and sequencers. Of course, Aaron knew this perfectly well. He just wanted to be sure I knew what I was up against; there was no doubt that, the decision made, he would support me all the way. As he says himself, 'John was the standard bearer [for genome studies], and that's why it had to be the worm.'

We had to go through the formal process of writing grant applications to the National Institutes of Health and the Medical Research Council. After twenty years at the LMB it was the first grant application I had ever written, and I was having to justify £1 million over three years to do our 1.5 megabases. Nor could I ignore the slightly chilling feeling that this was only 3 percent of the genome, and we were going to have to scale up. All this contributed to the 'prison door' effect I had first experienced at Syosset: I was going to have to stop playing and be a little more professional.

Initially we wrote our applications on the assumption that we would do the work the conventional way, using radioactive labels to tag the DNA fragments and film to record the sequence from the gels. We knew that there were some automated machines on the market but were initially skeptical about them; still, in order to cover ourselves, we applied for funds to buy one for each lab to experiment with. These machines attached different colored fluorescent tags to the DNA fragments instead of radioactive labels, and read them off automatically.

In September that year Bob Waterston came over to Cambridge so that we could write the grant application to the NIH together. Then he and I went off on a world tour that he had organized, to look at what was available in the labs where the new machines were on trial. There were really only two machines on offer, one made by Applied Biosystems Inc. (ABI), the company headed by Mike Hunkapiller that developed the inventions of Lee Hood's CalTech group, and one made by the Swedish company Pharmacia, which had the license from Wilhelm Ansorge at the European Molecular Biology Laboratory in Heidelberg. There was a third that had been developed by DuPont, which Sydney was using in his unit, but the company had never been able to get it to work as effectively as the others.

One of the labs we visited was at an outpost of the National Institute for Neurological Disorders and Stroke in Rockville, Maryland, where a researcher called Craig Venter was working on receptors for brain chemicals. He had been one of the first to take delivery of a prototype ABI sequencer, in February 1987. Since then he has focused all his energies on sequencing more and faster, moving into the private sector when he couldn't get what he wanted from publicly funded science. But all that came later. When we visited his lab Craig was keen to exploit the advantages of automated sequencing, having become frustrated at the years it had taken him to isolate and sequence the gene for a single brain protein using manual methods. He gave us the hard sell on the virtues of fluorescent sequencing, but it was not his lab that convinced us; we did not really see what was going on. The place that convinced us was a lab at Baylor College of Medicine in Houston, Texas, where Richard Gibbs (an Australian molecular geneticist who now heads the human genome sequencing center at Baylor) was putting all kinds of things through an ABI machine and getting good, clean data. We found it very impressive. Then we went with Alan Coulson to Heidelberg to talk to Wilhelm Ansorge, who was also getting good results with his own machine, which we could buy through Pharmacia. The two machines had different advantages and disadvantages, but we came home convinced that both were superior to the radioactive method because they instantly gave you the data in digital form. So we rewrote our grant proposal to include one ABI machine and one Pharmacia machine for each lab. And the funding agencies gave us everything we asked for. I still have the notification from the MRC, a classic missive. It was a really big grant for them, over £1 million for three years—and it came in the form of a hand-scrawled fax.

While we were in Heidelberg something else was decided. If you cross the River Neckar from the old town and head up the steep north side of the valley, you find yourself on the Philosophenweg. Presumably it was a traditional promenade for the academics of Heidelberg. You can follow it up through the forest until you come to a place where there's a pub and they keep wild boar in pens. And then you go back down to the Neckar and so home along the river-bank. Bob, Alan and I walked that way, and as we went we talked about what we had to do. I had quite expected that after seven years off Alan might be ready to go back to sequencing and to take charge of the project on our side. No way: the map was nowhere near finished and he would have a lot of work to do on an interface between it and the sequence. He also expressed the general principle that one should never go back. I returned from that walk in the woods feeling a mixture of resignation and excitement that I would have to learn a new skill.

I thought I had better teach myself sequencing the old-fashioned way before the machines arrived. I got the protocols from Alan and ran some radioactive gels (Alan had a good laugh at my first results), and started sequencing a cosmid. And that's what I did for the few months before we got an ABI machine. I personally was not mentally dependent on the machine. If it had been impossible to get the fluorescent readout going and none of this had worked, we could even have done the worm the old way, on film. At the time I was working with the LMB workshop and Amersham International to adapt the film reader we had built for mapping so that it could read DNA sequences. If we hadn't succeeded, perhaps some other form of direct readout would have taken over. Ever since Fred's invention of dideoxy sequencing, the improvement of technology in this field has been helpful but not crucial—rather like the development of the car since the internal combustion engine was invented. It's not the difference between being able to do nothing, and suddenly having a new technique—as it was with the lineage, which was absolutely dependent on Nomarski microscopy. Sequencing has been improved incrementally through machines, chemistry, enzymes and software, gradually allowing us to reduce costs and automate. Like the car, though, cumulatively it's come a long way. Although we couldn't have known this in advance, the way to make progress in sequencing has been to do the best you can with the technology you have available, not to spend years trying to invent a better technology (or, worse still, waiting for someone else to invent it).

A much bigger change in my life than starting sequencing was the business of taking on new people and running a group. Bob and I reckoned we needed about ten people each; I had never supervised more than one technician before, and hadn't made a very good job of that. With a technician who is entirely dependent on you, you have to come in in the morning and give them something to do. This used to make me sweat, thinking, 'I've got to give this person something to do.' I don't want to be thinking, 'What's so-and-so going to be doing in the morning?' I want them to know what they're going to be doing in the morning and looking forward to it. It wasn't easy for me to make the move to running a group. I had not learned to be a manager, or to delegate at all. I had never had to: until Alan came I just did everything, and then I worked happily as half of an equal partnership. As Rodger Staden once commented, 'The trouble with John is, he always wants to do everything himself.' The few post-docs I had, such as Judith Kimble and Jim Priess, who worked on the development of the worm embryo, were very strong and did things for themselves—Judith organized me rather than the other way round. It never occurred to me to think about how the numbers of people would grow. In so far as I thought about it at all, I thought there would just be more "John'n'Alans" around—a completely horizontal organization. It took me a while to realize that going into sequencing was going to lead to a big management structure, a very different way of organizing things from the research labs I was used to.

Some new space had been earmarked for us in the building adjacent to us in Hills Road. It had previously been occupied by the Neurochemical Pharmacology Unit, so we called it the Old Pharm, which I thought had a nice domestic ring to it. Michael Fuller, the LMB's indispensable laboratory superintendent, who always knew where to find that elusive piece of kit, was responsible for getting the space fitted out and equipped—and I appreciated more than ever before how vital his role was. But at first the whole group just piled into 6024. We stacked the sequencing machines vertically, and we each had a meter or so of bench. My entire office space consisted of another meter of table sticking out sideways with the phone on it. More like bedsit than open plan, but a great way to work.

We started advertising for staff in the press and by word of mouth, looking initially for graduates and Ph.D.s, and quickly acquired another half-dozen colleagues. I had no idea how everyone would fit in, and at first everyone did everything. This was the way the manual sequencing groups had worked, when techniques were developing rapidly and it was satisfying for graduates to learn all the skills. But the disadvantages of this method soon became clear: it took too long for each person to learn everything, and it was all rather erratic, with different things in the complex process going unexpectedly wrong, and difficulties in finding out what actually had gone wrong. So we started to sort out the jobs. The worst bottleneck had been making the libraries of worm DNA, so for the time being I took that on. This quickly started things moving, and had the added advantage that I could experiment with different ways of breaking up the clones and cloning the fragments in bacteria. The others all carried out the routine tasks of first growing the subclones, then reacting together the DNA templates, enzymes and nucleotides, and finally loading the fluorescently labelled products of these reactions on to the sequencing machines. They also performed the much less routine task of what we came to call finishing. This involves sitting at a computer screen analyzing the shotgun results: comparing the initial assemblies with the evidence of the raw data from the machines, represented on the screen by a set of four different colored traces (one for each base), then either editing on the basis of this visual evidence or performing additional reactions to get more data.

In the next wave of recruitment, as we expanded into the Old Pharm and bought more machines, we realized that it would be worth dividing up the work to make the best use of everyone's abilities. The existing staff, now skilled in sequencing, would become full-time 'finishers', as well as participating in technical development. We would recruit unskilled people, who would carry out the template preparation and reactions, but would also have the opportunity to train for finishing. This group would have no need of academic qualifications. We judged them on school achievements, interview and something by which I set great store: the pipetting test. I showed the candidates how to use a pipette—a hand-held tool for manipulating small volumes of liquid—and invited them to have a go. It's really simple, but the way in which a person goes about it gives an indication of their manual dexterity.

Our management principle was simple. You bring in people, find out what they do better than you do yourself, and hand that over. When there's a gap, you fill that yourself until that job can be handed on in its turn. And in that way things informally move forward.

Computer analysis was a vital part of the work, and we really needed someone on the team with the right skills. Richard Durbin, who had helped us out with the image analysis program during the mapping phase, had gone off to California on a postdoctoral fellowship, and John White thought it would be a very good thing if we tried to lure him back to work on the software side of sequencing. Richard is an extremely bright mathematician, but unlike many mathematicians he also likes making things work. So I asked him to join us, and he agreed. While he was in Cambridge Richard had become friendly with Jean Thierry-Mieg, the husband of a French biologist called Danielle Thierry-Mieg who had worked as a visitor in my lab for a couple of years. Jean, a great bear of a man who reminds me of the actor Gérard Depardieu, was a theoretical physicist, but was very keen to move into bioinformatics and work on the worm project (indeed, I once looked out of the window early one morning to find him sitting on the lawn waiting to start a discussion). Although the Thierry-Miegs had moved to Montpellier by the time Richard came back, Richard and Jean between them got to work and developed a new database program called ACeDB (A *C. elegans* Data Base) that provided a way of displaying the sequence, and the genetic map, and a bibliography of relevant articles on the genes. They distributed copies to all the other worm labs, so that everyone could have it sitting on the desk and update it as new data became available. It was a superb piece of work that set standards throughout the whole field, and went on to be used for many other genomes.

Bob also made some critical new appointments. He'd talked to some colleagues at St. Louis about making an appointment in computation jointly with them. While he was in England with me writing the grant application, one of them called saying they had found the ideal person in LaDeana Hillier, and could Bob agree long-distance to his part of the hire? So she joined without ever seeing the person she would be working for. She proved to be brilliant at developing the resources Bob needed to analyze the data that began to pour out. Then he hired Rick Wilson, who came from Lee Hood's lab at CalTech. Rick, whose background was in cellular immunology, had been testing the new sequencing machines on some clones of regions of the T-cell receptor gene, and held the record at the time for the length of a sequence from an organism more complex than a bacterium: 91,000 bases. It was very valuable for our collaboration to have someone like him who already had some familiarity with the technology, but who was a biologist at the same time. And for Bob it was a major boost to his request for supplementary funds to buy the sequencing machines to be able to say that Rick was joining.

> Rick knew the machine's limitations and what you had to do to get it to work. It took twenty-four samples, and at first we were running it once a day for fourteen or sixteen hours. We were so proud when we figured out a way to run it twice a day.

(For comparison, modern ABI sequencers run ninety-six samples at a time and can do eight runs a day—and some of the bigger genome labs have 100 or more of them in operation at once.)

Once the first fluorescence sequencing machines arrived, it became clear that we had to take control of the software. The machines worked well, but ABI wanted to keep control of the data analysis end of things by forcing their customers to use their own proprietary software. In order to finish a sequence properly in the way I described above, you must have easy access to the raw data in order to evaluate their quality from point to point. A good way of displaying the readout from the gel is as a set of colored traces on the screen. ABI's software produced a display, but not in a form that we could combine flexibly with Rodger Staden's assembly programs. It was inconvenient to use and slowed us down. I could not accept that we should be dependent on a commercial company for the handling and assembly of the data we were producing. The company even had ambitions to take control of the analysis of the sequence, which was ridiculous. I had a complete obsession with getting data out—I saw that as the bottleneck. There were an awful lot of people out there theorizing about genomes, so for the moment I didn't see that as our job. The best way to drive the science was to get the sequencing machines going, cheaper and faster, and get the data out so that all the theorists in the world could work on the interpretation.

So, one hot summer Sunday afternoon, I sat on the lawn at home with printouts spread out all around me and decrypted the ABI file that stored the trace data. I don't think it was deliberately encrypted; it was just constructed in a rather Christmas-tree-like fashion, which I needed to track from one point to another. I came in on Monday morning and said, 'Look, this is how we get the file data.' Within a very few days, Rodger and his group had written display software that showed the traces—and there we were. The St. Louis team joined in, and they all went on to decrypt more of the ABI files, so that we had complete freedom to design our own display and analysis systems. It transformed our productivity. Previously we'd only been able to get the traces as printouts, which we bound together in fat notebooks, infuriating to fast workers such as Rick Wilson.

> You'd sit down at the computer, and you'd have to flip through this stupid notebook until you found the trace that you wanted. And hopefully it would be in the right direction. The great idea was to figure out a way to get this on line, but ABI would not help us out. So John sat down and cracked it. That was a huge advance, really an important development. If we hadn't done that we'd have been way, way behind on the worm project.

ABI was not at all happy that we had done this. We had been negotiating towards the idea that they would sell us a key that would unlock the files, but it was quite clear that even then they would always have control and they could take it away again. There remained a real risk that they would re-encrypt the file in a way we couldn't get at; so we made sure that their other customers were aware of what was going on, and they did agree quite quickly to keep their formats public. We went on to become one of their biggest customers. I think I was the first to decrypt the files, but I'm not certain—there were others doing it at about the same time. I certainly feel that between us we did push ABI back a bit and denied to them complete control of this downstream software. It was my first experience of the kind of battle for control of information that I seem to have been fighting with commercial companies ever since: a foretaste of the much larger battles that would later surround the human genome.

Working with Bob and the rest of the St. Louis lab was a great experience. We compared notes constantly. We divided up the job by each starting at the same place on a chromosome and sequencing away from one another in opposite directions. That way we had only one overlap between the labs to worry about per chromosome. If it seemed like one lab had a particular problem covered, then the other left it to them. Problems that were easy to solve and didn't take too much effort could be tackled by both labs, just as it was worth each lab taking a different approach to solving the really hard ones. Yes, there was competition, particularly over how much sequence each side was producing; but it was a competition in which nothing was kept hidden. For example, early on Bob's lab had difficulty assembling cosmids.

> There were huge gaps—it just wasn't coming together. We eventually worked out it was due to how we were making the libraries—one of the extraction steps was partially melting the DNA. But Cambridge didn't have the same problem, so we had to work through our protocol with John. It turned out that we weren't doing it cold enough. We were trying to help each other as much as we could, but still trying to beat each other!

Being thousands of miles apart wasn't really a problem. We used e-mail a lot, and at some point Bob and I got into the habit of talking together on the phone once a week. Individual members of the two labs visited each other regularly. The highlight of the year was the annual lab meeting, when we took it in turns to host a visit from all the members of the other lab. These gatherings had a serious purpose—to see at first hand how the other group was working— but they're chiefly remembered for the social side. The Cambridge lab meetings typically involved punting, picnics and a healthy quantity of beer, a continuation of the tradition established by the worm group.

It was an extremely productive formula. After a couple of years we were far and away the biggest producers of genome sequence in the world. To our irritation we found ourselves having to respond to critics who simply refused to believe our production figures. Bob had a particularly unpleasant argument with one of the *E. coli* sequencers who had massively overestimated the capital costs of the project, simply because he refused to believe that Bob's lab ran the machines twice a day, seven days a week. A more serious question was whether people would believe what we were doing was worthwhile. We were getting flak from some members of the usually harmonious worm community who felt that we were taking resources away from them. Some people thought it was wrong to spend millions of dollars of research funds on work that was not directly going to solve problems in biology. What we had to do was to demonstrate that this was seedcorn, and that the money invested in the genome was money well spent.

A year after we began, at the 1991 worm meeting, some of the best labs were complaining. But two years further on, they found that people were pouring into the field, because they saw how easy it was to find genes. So the leaders of the labs suddenly saw that, far from reducing their grants, the genome project was actually increasing them, because it made everything more attractive to funding agencies. I began to notice that the worm grant proposals I was asked to referee all began to mention the genome.

What was perhaps more significant was that our project was proving the point that genome sequencing was worthwhile, that production went up and costs went down with experience and technology development, and that the clone-based approach was efficient and accurate. There was no doubt that, given the funding, the worm genome would be completed, and within a few years. And if two labs could do the worm for a few tens of millions of dollars, then the human was within reach too.

3 IN BUSINESS

I WAS STAYING IN THE PENTHOUSE SUITE IN THE CLAREMONT HOTEL in Berkeley, the white tower that you can see from all over San Francisco Bay. It has a private staircase up onto the roof, where you can stand and see the whole of Berkeley spread out below you. I was in Berkeley as part of a National Institutes of Health team that was reviewing the progress of Gerry Rubin's lab. Since leaving the LMB Gerry had made his name in *Drosophila* molecular genetics, and was now beginning to sequence the fly genome. Site visits didn't normally entail such extravagant accommodation, but on this occasion the Claremont was short of rooms and so I got this splendid suite as part of the NIH contract. It provided an appropriately sumptuous backdrop to what followed, as if the scene had been set up by a film producer.

That evening in January 1992 Frederick Bourke, a wealthy investor who had made a fortune in leather goods, came up to see me. It was our third meeting, and things were not going the way he expected. At one point he remarked that 'dealing with scientists is like being a sheepdog rounding up sheep.' He wanted Bob Waterston and me to head the commercial sequencing organization he was hoping to establish in Seattle. As on previous occasions I was non-committal, but this time he probably saw in my eyes that I wasn't going to accept. He said 'John, I do hope that this isn't going to do you any damage.' I said it would do me no damage at all. There was no harm in being known as somebody who was wanted in America.

After Bourke had gone, my daughter Ingrid and her boyfriend Paul Pavlidis came over to the hotel with friends. Ingrid had come to Berkeley in 1990 to do a Ph.D. in developmental biology. She had been so eager to go to California, I used to tease her about being imprinted on the land of her birth. We went up on the roof and smoked a little, as one does in Berkeley, and looked out over the bay. I remember thinking that it was another significant moment. But I didn't know that within months, and as a direct consequence of this slightly surreal episode, Jim Watson would have lost his job heading the NIH genome program and I would be committed—that prison door again—to heading the laboratory that would ultimately complete the sequence of one-third of the human genome.

Given our commitment to the public domain, the fact that Bob and I were even considering heading up a commercial organization needs a bit of explanation. It was all because of the success of the pilot worm sequencing project, launched in Jim's office two and a half years previously at Cold Spring Harbor. By the end of the second year it was clear that we were going to meet the targets set for the three-year pilot project. I had never doubted that we should then scale up from doing 3 percent of the worm genome to doing the other 97 percent, and Bob was just as confident. But we were very unclear in our own minds whether either of the funding agencies on which we depended was going to come up with the money we would need. Our doubts arose partly because of the flak we were getting at this stage from those who doubted the utility of the worm genome, but mainly because the sums involved were going to be so large: we had slashed the cost per base, but we would still need in total ten times as much as we had had before. We weren't very confident that our funders would think we were worth that kind of expense, so we were very open to exploring other possibilities. We had never before had this sort of responsibility, both to our project and to the teams working on it. Bob and I were both worried about it. We certainly planned to put in further grant proposals; we were just concerned that we might not get the money. For the next stage there would be no more funds coming to the U.K. from the United States government. Jim made this absolutely clear: for the pilot phase he could do a deal, for the production phase, no. If NIH was going to put in half, then the MRC had to put in half. And now the MRC was really challenged: they would have to find about £10 million.

Then, in 1991, with a year to go before our pilot funding ran out (Bob's had come later, so he had a little longer), Lee Hood phoned up out of the blue and said, 'I have a proposition for you. I want to start a sequencing organization, and I want you and Bob to come and lead it.' Lee is a very remarkable guy, and a great networker. The fact that we were sequencing so efficiently owed a great deal to the ingenuity of his CalTech lab in starting the development of the automated sequencers, followed by the founding of ABI under Mike Hunkapiller to turn them into a commercial product. Lee had just agreed to move to the University of Washington, Seattle, to start an academic biotechnology department with money from Bill Gates. Now, he told us, he was proposing to start a commercial sequencing operation alongside the department, and he had persuaded Rick Bourke to put up the capital. He wasn't offering us academic positions, but we would be somehow affiliated to the university, and he implied that we would be able to complete the worm genome sequence.

If it had been anyone but Lee, we might not even have considered it. But he was someone we respected, who was running a good academic lab. We were intrigued enough—and doubtful enough about our chances of getting funded by our government funding agencies, or indeed anyone else—to fly over there and have a couple of meetings with them all.

Initially, Bob and I were much taken with the idea. We would have been in the same city, working side by side. 'That was a very attractive part of it,' says Bob. 'Lee Hood was going to be there, with the technology push that he had. He was also trying to attract Maynard Olson to his group [Maynard moved there soon afterwards], and you could see the attraction of setting up a real local powerhouse.' We were seriously trying to think, 'What does Lee want? Do we want to do this?' Bourke was offering us salaries that were enormous by my standards, and of course stock options; and, more to the point, if we wanted to get into large-scale sequencing this might be one way to do it. And then, little by little, we began to realize that sequencing the worm wasn't on Bourke's agenda at all. He began to complain about how 'all scientists came with baggage.' And when we began actually to write down what we were going to get out of the deal, he started saying things like, 'How many machines can you bring with you?' More seriously, we differed on the question of releasing the sequence data into the public domain. Bourke's initial idea was to patent the sequence. We said that wasn't necessary, we could just introduce a delay of a few months before releasing the sequence that would give a research team time to find genuinely useful regions. He didn't buy it.

It gradually became apparent that we were simply going to be scientific directors of a commercial sequencing organization, and there was less and less possibility that we would get the worm done at all. It was quite clear that all we would get would be premises—which we already had; if we wanted to do any of our own research, we would have to go out and get grants in the usual way. And with both of us in the United States, we would have to go after NIH money jointly; we would lose the international collaboration, with the MRC as equal partner.

By the time of my meeting with Bourke in the Claremont Hotel, we were close to deciding against the scheme. Soon afterwards we set up a conference call and Bob and I told him it wasn't going to happen. I was never in any difficulty about the decision. One has one life and one life only, and one does what one really wants to do. I don't see the point of making money for its own sake.

Meanwhile Jim Watson got wind of the negotiations with Lee Hood and Rick Bourke, and as far as he was concerned, Bourke was trying to pinch his project.

We were worried that Bourke would recruit John and Bob, because the word was out in the fall of 1991 that they were the only success- ful sequencers at that time.

It was true that, for a variety of reasons, the other pilot projects that had been funded in the first round of genome project grants, on *E. coli*, yeast, *Drosophila* and parts of some human chromosomes, were not producing anything like as much sequence: the worm was, in effect, the only game in town. Jim arranged to meet Bourke, and was incensed to find him bent, as Jim saw it, on destroying the NIH genome program for a private company. As well as wanting to retain the most successful sequencing teams, Jim was very anxious not to lose the international aspect of the sequencing project, which at the time was basically the worm genomics group and the MRC.

I thought the fact that you had two countries coming together was great, because if one put up money the other had to, whereas if there was only one, if it cut back you couldn't say there was someone else doing it. So I fought Bourke.

Jim immediately began to contact everyone he could think of to prevent us from joining Bourke. But first he rang Bob Waterston. At this point, in mid-January 1992, we had still not finally made up our minds. After talking to Jim, Bob realized that we could turn the situation to our advantage—'use it to open the spigot a little bit,' as he puts it. He wrote to Jim to let him know in uncompromising language that our basic problem was lack of funds. While the NIH was apparently sitting on its hands, new genome efforts were beginning in France and the United States with private or charitable support. 'We did not start this project to be outdone by second-rate players with more resources,' wrote Bob. But he emphasized our commitment to keeping genome work in the public domain. 'We, of course, have some remaining misgivings about a venture so foreign to what either of us has done before ... the genome project cannot prosper if it is viewed simply as a way of making its creators rich.'

The same day that Jim called Bob, he also spoke to Aaron Klug at the LMB, poured out his 'annoyance and rage' at Bourke's venture, and told him that he would be in England at the end of January on his way to the World Economic Forum in Davos, Switzerland. On the way he called in to see Dai Rees, secretary of the MRC, to tell him how concerned he was about the possibility of my being lured to the United States. Then he came over to Cambridge to talk to me. I confirmed what he had already heard from Bob—that we had considered Bourke's offer only because we couldn't be sure we would get the money to complete our project. But by this time we were close to saying no, and he left reassured that we weren't on the point of deserting him.

Jim probably didn't realize that, by rushing around making a stink on our behalf, he was locking the stable door when the horse was still firmly inside and had bolted itself in. I don't know why we hadn't taken him into our confidence much earlier. As it was, his intervention added fuel to a fire that had already been lit underneath him at the NIH.

James Wyngaarden had been succeeded as head of the agency by Bernardine Healy, a career scientific administrator. She and Jim did not see eye to eye. The main bone of contention between them at this time was the patenting of gene sequences. The previous summer, the NIH had filed patents on several hundred fragments of genes, known as expressed sequence tags or ESTs (see p. 105). The ESTs had been generated in the lab of Craig Venter, in the course of research that was independent of Watson's genome center. ESTs are convenient labels that help to identify genes, but do not in themselves tell you anything about the function of a gene (unless you can match them to genes in the same or other species that already have known functions).

Patents (or so I had always believed) are designed to protect inventions. There are three essential criteria for an invention: it has to be novel (no-one has published the idea before), useful (in that it could be developed for commercial or other uses) and non-obvious. The ESTs met none of these criteria. There was no 'invention' involved in finding them, so how could they be patentable? Yet the 1991 patent application claimed exclusive rights not only to the ESTs, but to the whole genes they represented and even the proteins encoded by these genes. It was crazy. But Healy had arrived determined to push the commercial development of scientific discoveries. There had been no proper debate, nationally or internationally, about where the line should be drawn on patents for genetic discoveries, but the NIH lawyers decided to play safe by putting in patent applications on Craig's ESTs. That way, if scientists in other countries started patenting genes, they would have a prior claim.

The decision to apply for the patents involved a number of different branches of the NIH bureaucracy, but the genome research center and its head, Jim Watson, were told it was happening rather than asked for their opinion. When Craig announced at a public briefing on genome research for the United States Senator Pete Domenici in July 1991 that the patents had been filed, Watson burst out that the move was 'sheer lunacy', and said he would be 'horrified' if it were true that random bits of sequence could be patented.

He argued that there was no invention involved, asserting that the automation of sequencing meant that the work could be done by 'virtually any monkey'—a typically unguarded remark that was probably more insulting to Craig and his colleagues than was intended. Jim was extremely concerned that premature patenting of sequences whose function was unknown could undermine the delicate structure of national and international collaboration that he saw as essential if the benefits of the genome project were to be fully realized. Healy, on the other hand, believed that without patents on the sequences potential licensees from United States industry would have no interest in pursuing commercial development of the discoveries. It was not a matter on which the two of them could agree, and their dispute continued through the medium of public pronouncements on either side rather than face-to-face discussion.

Watson eventually agreed not to criticize NIH policy publicly, but others in the academic community could speak for themselves. The scientific advisory committee on the Human Genome Project, chaired by Paul Berg from Stanford University, said it was 'unanimous in deploring the decision to seek such patents ... we believe such claims are inappropriate and deleterious to science.' Like Jim, the committee feared that the claims would set off an international 'patent race' that would destroy the collaborative framework of the genome project. HUGO had already made a similar statement and Walter Bodmer, now HUGO's president, confirmed that if the United States agency persisted with its patents then the U.K. would have no alternative but to follow suit. Despite these protests, in February 1992 the NIH added a further 2,375 ESTs to the patent application, although this time they dropped their claim to the proteins encoded by the genes they represented. Their action set off a fresh round of criticism. Berg spoke for many when he said, 'It makes a mockery of what most people feel is the right way to do the Genome Project.'

Into this volatile atmosphere came a letter from Bourke to Healy, backed up by a phone call from Lee Hood, accusing Jim of acting against the interests of American industry by interfering in Bourke's attempt to recruit Bob and me. 'I was not anti-American,' says Jim now, 'but I was certainly anti-Bourke because I didn't want the one success that we had broken up and in the hands of private industry.' Healy used Bourke's concerns as an excuse to set up an investigation into Jim's financial holdings on the grounds of suspected conflict of interest. Nothing was found that he had not previously declared, but Healy never publicly exonerated him. Jim felt he was left with no option. He resigned in April 1992, but today he is in no doubt that he was effectively fired.

> It turned out I always had an illegal job. I should never have been head of Cold Spring Harbor while passing out money from NIH. But at the time [when he first came to the genome office] I don't think anyone would have quit their job to run the Human Genome Project—at least no one of real calibre.

Jim returned to Cold Spring Harbor declaring his continued support for the project under whoever turned out to be his successor. But the HGP had lost a hugely valuable asset—a bold and single-minded leader who would battle publicly for what he believed in, and one who had made the project truly international. At least Jim had established the principle that the project should be led by a scientist and not a career administrator. A year after his resignation Healy appointed Francis Collins, the University of Michigan geneticist who had been involved in finding several disease genes including that for cystic fibrosis, to succeed Jim. In the midst of a productive phase of his own research career, Francis was a reluctant recruit, but agreed to move to the NIH campus at Bethesda if the job could include facilities for his team and other new genetics investigators to pursue in-house research. The National Center for Human Genome Research later became the National Human Genome

Research Institute (NHGRI), which both conducted its own research and dispensed funds to other genome centers. Francis, a tall, spare figure and a devout Christian, was to remain in the hot seat—the NHGRI became the de facto coordinating center for the international genome project—through to the present day, deploying his gentle Southern charm in an effort to build consensus among the often hostile factions that had an interest in the human genome.

The controversy about the EST patents rumbled on for another two years. The United States Patent Office roundly rejected the first application in August 1992, but the NIH appealed against the rejection and applied for patents on another 4,448 ESTs. After the election of Bill Clinton as United States President in November 1992 Healy resigned (the NIH head is a political appointment and she was a Republican appointee). Her replacement, the highly respected cancer researcher and Nobel prizewinner Harold Varmus, decided not to pursue the patent issue any further and withdrew all outstanding applications in early 1994. The question of what constitutes an acceptable patent on genetic material is still not fully resolved, and arguments sway to and fro. The United States Patent Office in particular has granted thousands of patents on gene sequences, but anyone with the funds to challenge these patents in the courts can do so. Not until 2000 did the Patent Office produce a set of guidelines that tightened up the definition of 'utility' to prevent people giving uses as vague as 'a gene probe' in their applications. It is still permissible to patent a gene sequence as long as you can show how it might be used to diagnose diseases, for example. In the end the issues are being decided not on principled grounds, but according to which side has the most money to spend on lawyers. One of the aims of the Human Genome Project has been to 'raise the bar' by making as much genome information as possible universally available in the public domain and therefore unpatentable.

How much the Bourke episode contributed to Jim's departure from his key position in the genome project is hard to tell—the patent issue was probably the more serious source of conflict between him and Healy. But although from his own point of view Jim risked more than he needed to in order to keep Bob and me where we were, his intervention made a material difference to the fortunes of both the Cambridge and the St. Louis worm sequencing groups. Jim made the MRC realize how much the U.K. stood to lose if it failed to fund genome sequencing at a substantial level. After his visit, Aaron Klug immediately got Dai Rees's permission to contact Bridget Ogilvie, the recently appointed director of the Wellcome Trust.

The Trust is a charity that was set up by Sir Henry Wellcome on his death in 1936. He placed the entire share capital of his successful pharmaceutical company, the Wellcome Foundation (later Wellcome plc, then Glaxo Wellcome, now absorbed into Glaxo SmithKline) in the hands of trustees, and directed them to use the income from the shares to fund scientific research for the benefit of human and animal health. In the mid-1980s, following concern that the company's performance had reached a plateau, the trustees began to sell off their shares in order to widen the asset base and safeguard the substantial contributions they were making to research. At about the same time, there was a dramatic upturn in the company's fortunes following the introduction of the anti-HIV drug AZT. After the Trust sold a second tranche of shares in 1992 it became the wealthiest medical research charity in the world. Bridget Ogilvie, an Australian scientist who had worked for the Trust since 1979 after a productive research career investigating immune responses to parasitic nematodes, became its director in October 1991. Within a year her budget had more than doubled, from £91 million to £200 million (today it is more than twice this figure), all of which she had to spend. 'To spend money on that scale sensibly is not easy!' she says.

The MRC's approach to the Wellcome Trust could not have been more timely. Although relations between the Trust and the MRC had been strained in the past, Bridget (whose own research career had been conducted mainly in an MRC institute) was enthusiastic about entering into some kind of partnership.

> The first idea was that we would each put in £2 million, just to keep John. But I immediately saw that because of the sale of the shares, we had this massive amount of money and could do something more.

Very quickly it became apparent that the Trust was not interested in subsidizing our work at the LMB, but was prepared to think about spending very considerable sums on a much larger genome sequencing effort that would not only provide space for the worm sequencing group, but would also embrace human DNA. The MRC was put on its mettle—it wasn't going to allow the Wellcome Trust to steal away its project. In June the following year it came through with a grant of £10 million over five years to complete the worm sequence—a wonderful vote of confidence in the project that meant so much to me.

The year of 1992 was completely insane. It was one of those moments in one's life when one feels swept along like a leaf on the stream. First there was Bourke, and a burst of comment about the episode in the British press, which perhaps helped to propel succeeding events. Then I got sucked into negotiations with the Wellcome Trust for building a new sequencing center, of which I would be director, to make an attack on the human genome. In the spring Aaron Klug dragged me down to London to a meeting with the Wellcome Trust's Genetics Advisory Group. The group had been set up only the year before to develop a strategy for the Trust on genome research. At Aaron's instigation, I had hastily written a four-page briefing document, plus a few tables on the running costs, which I put at just over £10 million for the first two years. In it I described our progress on the worm genome and set out how this experience could be transferred to the human. I concluded, 'Surely now is the time to begin sequencing on a large scale, rather than, as has been argued, waiting upon the emergence of still better instrumentation and ingenious genetic tricks... To begin now gives Britain the opportunity to show the way.'

In the meeting Aaron adroitly made the case that large-scale sequencing, or megabase sequencing as we called it, was going to change things totally in the search for disease genes. He took the example of the gene for Huntington's disease, which had been linked to a chromosome almost a decade before but was proving extremely difficult to locate definitively. Trying to get ever closer to the gene through traditional genetics, looking for linkage with other markers, took a lot of time and a certain amount of luck. On the basis of what we were doing in the worm, Aaron argued that once you had got within a few million bases you could stop doing linkage studies and sequence selected mapped clones instead. 'That was the argument,' he says; 'that you would be able to get results on the way—even before you had the whole sequence you would be able to home in on the gene of interest.'

Everything began to fall into place with astonishing speed. In March 1992 Bridget Ogilvie appointed one of the senior Wellcome Trust administrators, Michael Morgan, to look into the viability of a Trust sequencing initiative. Meanwhile we had to develop my briefing note into a formal proposal to cover the first five years. We were talking about a highly organized operation several times the size of our existing worm sequencing group—bigger, indeed, than any other sequencing center in the world at the time. The proposal was for a grant of between £40 million and £50 million. (And to think that less than three years before I had gone weak at the knees at the sight of a £1 million grant!) Considering the sum of money at stake, our proposal was really a very tiny document. It included a table that I made up almost as a joke one day, that showed how in five years we were going to finish the worm, *and* complete the sequence of baker's yeast (*Saccharomyces cerevisiae*) alongside the international consortium that had recently started, *and* sequence the first 40 million bases of the human. Actually, we very nearly did it: we were just about six months late all through, which was the difference between the optimistic nominal start date and our actual move to the new site. Anyway, it worked. The governors considered the proposal, and agreed to fund it, just like that. 'One of the pressures was to do it quickly,' says Bridget Ogilvie.

In the summer of 1992, anticipating (correctly) that there wouldn't be another proper holiday for a long time, Daphne and I took off for Glacier Park in Montana. Instead of going direct, we flew to Seattle and hired a car there. There was a little nostalgia in looking around Seattle on the way, this beautiful place where we had almost gone to live. Glacier was perfect, with its lakes and mountains and long footpaths and bears. Ingrid came to join us halfway through, riding the long train that snakes slowly up from the coast, and together we walked the two-day trail over the continental divide. It was a welcome interlude in a chain of events that was sweeping me along, to what end I could not begin to predict.

During the holiday I put in one or two calls to Michael Morgan, to see how things were going. One of our first tasks was to find a building to put the new center in. Over the phone he told me about various possibilities that had come up, including a disused chicken research institute—he joked that the big shed looked like a possibility for the sequencing machines, even though it was too low for a person to stand upright. When I got back from holiday Michael and I spent a few weeks going round looking at office blocks in Cambridge, without much success. We also briefly considered Edinburgh; it was a growth area for hi-tech industry, and would have been a wonderful place to go—like Seattle, with hills and water. But regretfully I turned my back on such frivolous considerations: it would have delayed us to move so far because not everyone would be able to go, and it seemed wise for reasons of continuity to stay in touch with our base in Cambridge.

Then we heard from Michael Fuller, the lab superintendent at the LMB, that a property developer was letting an estate in the village of Hinxton, about nine miles south of Cambridge, which had previously belonged to the metallurgy and engineering firm Tube Investments. The estate included the rather run-down Hinxton Hall, a small stately home dating from the eighteenth century, plus 55 acres of parkland and a range of brick laboratory buildings that had been built as a research center for TI. The developer had intended to knock down the labs and build a business park, and had planning consent for 125,000 square feet of new buildings. But finding no takers for his plan in the early 1990s property slump, he put the whole estate back on the market. We went to have a look, and were charmed right away. The estate is set among open fields, and the river Cam runs along the boundary on its way to Cambridge. Michael and I saw that the labs could be converted into a genome research facility in a matter of months. There was ample space to get going, though the situation out of town and 8 miles from the LMB caused some heart-searching. The Trust agreed to let us take a one-year lease on the estate and carry out the conversion. But the refurbishment of the Tube Investments buildings had barely started when another major development took place that materially affected the future of the project.

During the winter of 1992–3 I was on a committee talking about the future of the DNA Data Library of the European Molecular Biology Laboratory, an international research organization funded by a consortium of countries. The Data Library, one of three international repositories for sequence data, along with GenBank and the DNA Data Bank of Japan, was still housed in the European Molecular Biology Laboratory's headquarters in Heidelberg. The committee was discussing the idea of developing not just a growing sequence repository but a European Bioinformatics Institute. The institute would store data but also develop the tools to do research on it in a computational way. The Germans assumed the new institute would also be in Heidelberg; they had planning permission, and were all set to build it there.

Michael Ashburner, the fly geneticist, was also on this committee. He took it into his head to persuade the Data Library's director Graham Cameron that we should get the whole thing to move to the U.K., and I backed him. The Wellcome Trust and MRC were enthusiastic about the idea, and they moved quickly to establish the principle that there should be an open competition for the institute, rather than letting it go to Germany by default. We then began to put a bid together. Immediately we hit a snag: neither Wellcome nor the MRC owned a suitable site. Michael Morgan proposed Hinxton, although the Trust did not actually own it at the time. The Wellcome Trust governors solved that problem by agreeing, at very short notice, to buy the whole estate. So out of the blue these cheeky Brits put in a proposal to host the European Bioinformatics Institute and won, on the basis that the institute should move in next door to what was going to be Europe's largest sequencing center.

That coup determined us to stay at Hinxton. Although the idea of the lab being there on a temporary basis was established, we had not previously decided that it was going to be permanent. The estate cost 'two or three million' pounds to buy; it was a perfectly good investment, with the property market in such a state of decline. But winning the bid for the bioinformatics institute prompted Wellcome to make a much bigger investment in the site. By the time we moved to the refurbished labs in April 1993, there were already plans to replace them with a sleek, modern laboratory surrounded with pools and lawns, and to restore Hinxton Hall as a conference center. The institute would go up next door, and in 1994 the MRC also moved its Human Genome Mapping Project Resource Centre to the site, now known as the Wellcome Trust Genome Campus.

We decided to name our new lab the Sanger Centre, in honor of

Fred Sanger, who had first made DNA sequencing a practical possibility. It fell to me to phone Fred up and ask his permission to name the place after him. I dialled his number in some trepidation, not knowing what his reaction would be, for Fred famously shuns publicity. He accepted immediately, but on one condition. 'It had better be good,' he said, and once again I heard the door closing, as it had at Syosset. In October 1993, a few months after we had moved in, we held an enjoyable opening ceremony at which Fred himself unveiled a commemorative plaque and we presented him with a security pass to the building so that he could visit whenever he liked.

Money and buildings were all very well, but the success of the Sanger Centre was going to depend on the people we found to staff it. From the moment it was first mooted, I realized that I would need an excellent administrator, never having run anything like this before—though it's not quite true to say I had never held any position of responsibility other than running the worm sequencing team. Not long before, Aaron had made me head of a new section at the LMB called Genome Studies. When Sydney left in 1986, he left behind a loose grouping called the Director's Division, and at least as a stopgap Aaron felt the best thing was to keep it together as an entity with me as head. 'It seemed very clear to me that although [John] eschewed responsibility he was actually taking it on willy nilly,' says Aaron. But even so, heading Genome Studies, which consisted for the most part of independent scientists, was not going to be the same as running a genome sequencing operation employing dozens of people. And, although I probably didn't quite realize it at the time, I was going to be directly responsible for the day-to-day budget in a way that I'd never previously experienced. I was going to need someone with real administrative skills.

Jane Rogers had moved from scientific research in Cambridge to scientific administration at the MRC head office in London. But she still lived in Cambridge and, with a young son to look after, found the commuting difficult. So she had asked if there was any possibility of being transferred back to Cambridge, and they sent her to talk to me. Alan Coulson and I interviewed her in my sitting room at home, an experience she found 'bizarre ... They plied me with sherry and asked if I could convert an office block into a lab,' she recalls. It seemed right straight away, and was to be the first of many interviews I would share with Jane, though thereafter we would be on the same side. (I find that usually after five minutes you know whether you can work with someone or not; from then on it's either real fun as you explore one another's thoughts or else you're trying to get out of the room as fast as possible.) I told the MRC that we were keen to employ Jane, and around the middle of 1992 she was seconded to us for two or three days a week, initially still in the lab at the LMB. The first thing we had to do was to write the formal proposal to the Trust that would lay the foundations for our as yet unnamed genome center. I tried to write something, dropping drafts through Jane's door as I cycled home in the evening for her to work on by the morning. The only grant proposal I had ever written before was the worm sequencing proposal, and Bob had had a lot to do with that, but through her experience at MRC head office Jane knew exactly what funding bodies were looking for. 'I knew how to write down what was needed and make it look plausible,' she says, and she played a big part in sorting out the budget pages so that they added up to the right amount.

The next question—of course, all these discussions were going on in parallel—was which scientists would move to the new center. It was to be a joint enterprise between the Wellcome Trust and the MRC, but it was clear from the start that in financial terms the partnership was unequal. What the MRC could bring to the table, in addition to their funding of the worm sequence, was the expertise they had nurtured at the LMB. Alan and I and our worm sequencing team would go, obviously. And Bart Barrell, who had sequenced more genomes than anyone and was now working on yeast, jumped at the chance to come.

> It was the obvious thing to do. If the funding from Wellcome hadn't happened, [the yeast genome sequence] would not have gone very fast at the LMB. We just had the same budget that we'd always had; every year it got smaller and we made it go further and crammed more people into less space, although as the technology got faster we could do more. So there wasn't a choice—you said right, if that's where sequencing is going, then I'm going too.

Then we needed people who could manage the bioinformatics side, writing the software to manage the flow of information, analyze it and display the results. Initially we expected that Rodger Staden would join us. For years he had written the sequencing software used at LMB, at first on his own and then leading a growing group, and the Staden package is widely used. He eventually decided against leaving the LMB, but we have continued to collaborate closely. Richard Durbin, who had written ACeDB with Jean Thierry-Mieg to handle the worm sequence, did move across to head the new center's bioinformatics group, and later he became deputy director.

If we were going to work on human DNA, we needed someone who knew something about it. All DNA is alike in that it all consists of the same As, Ts, Cs and Gs. But compared with the worm, for example, human DNA presents particular difficulties for the sequencer. One problem is the high proportion—more than 98 percent—of non-coding regions; the other is the fact that as much as half of the non-coding DNA is in the form of repeats. These repeats make assembling the DNA fragments read by the machines a bit like doing a jigsaw puzzle in which half the pieces depict either grass or sky. And, apart from the technical difficulties of reading human DNA, there was an establishment of human geneticists all with interests in particular human disease genes, a world about which I knew little. We needed someone with a foot in the human genetics camp, but who was also interested in human DNA for its own sake.

Right from the start I had in mind David Bentley, who at the time was working in the genetics department at Guy's Hospital in London. David had worked on a detailed analysis of the mutations in the Factor IX gene, a gene on the X chromosome that causes a form of hemophilia. But he was also interested in mapping larger genomic regions, and clearly understood that in the long run genomic mapping and sequencing was going to make the search for disease genes a whole lot easier.

> We tried to jump from the Factor IX gene to something that we thought was about a hundred thousand bases away, and we just couldn't do it. Suddenly the whole story started to illustrate quite clearly how we needed to get a lot more of the genes—we needed to get all the genes—in order to do more than just scratch the surface.

Earlier, David had formed his own collaboration with Ian Dunham and Eric Green in Maynard Olson's group at St. Louis, putting together a YAC map of the muscular dystrophy region of the X chromosome. After this Ian, a British scientist who had been doing a post-doc with Maynard, came back and joined David at Guy's. They embarked on a map of the whole of chromosome 22, one of the smallest human chromosomes.

Throughout this period David used to come and visit Alan and me at the LMB to talk about fingerprinting, and Richard to talk about ACeDB, which Richard had adapted to handle human data. 'I always enjoyed going up there,' he says. 'It was a breath of fresh air to talk technology and not be worried about genetic targets.' So when I began to think about who we needed to look after the human side of things, David seemed the obvious candidate. Inadvertently, I picked a rather awkward moment to approach him. He was in a meeting with his head of department, Martin Bobrow, when my call came through. Oblivious to the delicacy of his situation, I said, 'I'll come straight to the point. How would you like to come and join me in setting up an institute?' David nearly fell off his chair with surprise, and was left struggling to find a suitably diplomatic reply. But very soon he accepted.

> I was deeply integrated into the Guy's unit. But life got turned on its head by that phone call. It was clear that the time to scale up the genome was now, but until the Sanger Centre came along there weren't enough funds to do more than continue painstakingly going along bits of chromosomes, focusing on a rather genetic approach.

David brought with him his own project on the X chromosome, as well as Ian Dunham and the chromosome 22 mapping effort, so that from the start we had flagship projects on the human genome.

I thought the team of myself, Alan, Jane, Richard, Bart and David would be all we needed to manage the various aspects of the new center's operations. But the Wellcome Trust required us to have a chartered accountant to look after the financial and legal side. That was a foreign move for all of us—the MRC would never have made such an appointment, because these matters are handled by head office. But, because we were funded jointly by the Wellcome Trust and the MRC, a management company called Genome Research Limited was to be set up and would require a qualified company secretary. So a smart-looking advertisement went out in *The Times*, and the Trust recruited Murray Cairns, a genial Scot who had been made redundant after many years in management with Bass, as a result of the restructuring in the brewing industry. In order to make clear the respective responsibilities of Jane and Murray, she was called the Scientific Administrator and he the Head of Corporate Services. (Jane later handed over the scientific administration to Christine Rees, and became Head of Sequencing.) We worried at first that Murray might cramp our style by being too bureaucratic, and indeed I heard much later that the Trust hoped he would keep an eye on us as their watchdog. But despite his initial indoctrination he quickly 'went native' and became a doughty champion on our behalf against the bean-counting element at the Trust's headquarters in Euston Road. He also made us aware of our shortcomings as managers, and gradually helped us to see that, in an operation that was both a factory and a laboratory, it was essential to acquire management skills. Over the years Murray and I have spent more and more time together (particularly in the Red Lion at Hinxton) and learned a lot from each other.

This, then, was the team—Alan, Richard, Jane, David, Bart, Murray and me—that became the board of management (known as the BoM) of the new center. As I write, Murray and I are the only ones of the original team who have retired—all the others are still in place, a testament to their commitment and the excellent team spirit they developed.

In the first year after our move to Hinxton the staff grew from fifteen to eighty, and it continued to grow as the years went by. Jane and I undertook the initial recruiting of the staff we needed to run the sequencing. We ran our first advertisements in the *Cambridge Evening News*, and got people ranging from school leavers to a graduate in philosophy. According to Jane I had shocked the LMB by recruiting people who had no academic background. She accuses me of having favored women who'd worked as barmaids—I was certainly more interested in evidence of practical competence than paper qualifications.

I also had to get used to people leaving. At first I found it hard not to take this personally—how could this colleague not want to see the thing through as I wanted to? Then I realized two things. One was that the organization, as well as the people, was changing and developing very rapidly. A person who made a good fit at one stage was not necessarily a good fit later. The turnover was healthy and allowed evolution. The second thing was that we were actually training and educating people, just like a university, and so making a practical contribution to the needs of industry—U.K. industry in particular—even though we could not sell them anything.

I learned to share everything with the BoM and to pass on as much as possible; there was so much to be done that you had to delegate part of the planning, let alone the implementation of the plan by other people. I saw my role as that of a chairman. I vaguely assumed that someone would let me know if there was a problem, and didn't at first see the need for anything more formal in the way of management. But after a bit Murray persuaded us to go for some management training, and that was a real eye-opener. All the members of the BoM went away for the weekend to a hotel in Rutland for an intensive course with someone from a professional management training company. The first thing he did was present us with the results of a survey he had carried out among the hundred or so staff, which revealed that they had a pretty low opinion of our management skills. Some of them were quite unhappy—remarks about piss-ups and breweries were the milder ones. I was fairly abashed; I had no idea it was so bad. Although there was a certain amount of hilarity over the practical team exercises we had to do on the rest of the course, we did realize that we had something to learn. After that weekend we developed a more formal management structure throughout the whole organization, and gradually increased the number of levels in the hierarchy so that there was a career progression for people who did well. The other important outcome was that we realized how valuable it was to get together for a weekend like that, and thereafter we had a retreat every year—not for more management training, but just to discuss the scientific policy of the center and reflect on what we were doing.

The Sanger Centre was a totally different environment from the one in which I had worked until 1989, quietly putting together the worm cell lineage and then the worm genome map. Indeed, some of my oldest colleagues and friends could not hide their astonishment that I should have ended up in such a position. I remember standing with Bob Horvitz at the window of the old lecture room a year or so after we moved to Hinxton, looking at the enormous hole in the ground that was going to be our new building. 'John,' he asked me, 'do you really know what you're doing?' I replied that I realized that it was a big change, but that I thought it was important; so I was prepared to do whatever was necessary to make it happen. For myself, I didn't see it as a permanent change of direction, just as a means to an end. I always believed that once it was up and running I would be able to go back to doing something more sensible. I had forgotten what Alan Coulson had said during that walk in the woods near Heidelberg—you can never go back.

In many respects, things went on as before. Most importantly, the collaboration with St. Louis not only survived the change, but if anything was strengthened. Six months after we got our £10 million from the MRC for the worm sequence, the NIH also came through with funding for a scale-up in Bob Waterston's lab. Bob is in no doubt that the two events were connected: 'When it was clear that the Wellcome Trust and the MRC were going to create this larger venue for sequencing over there,' he says, 'it made the NIH funders much more receptive over here.' At the same time he also gained more space, establishing the Washington University Genome Sequencing Center on the fourth floor of an office building that the university had recently acquired.

By the end of 1993 our collaboration on the worm was not only the most productive genome sequencing operation in the world, but the biggest. We had easily exceeded our target of generating 3 megabases of finished sequence in our first three years. On top of that, Bart's group was churning out yeast sequence, and David was lining up some of our first human cosmids. But as we began to establish our reputation, the first signs had appeared of a belief that private enterprise could do as well or better than the public project, and could do it quicker and more cheaply. The first in the field was Craig Venter. Frustrated at the lack of support within the NIH for his venture into genome research, he left in July 1992 to set up his own privately funded, non-profit-making genome center, The Institute for Genomic Research (TIGR), in Rockville, Maryland.

Craig had joined one of the in-house NIH labs in 1984, and spent years looking for the gene for the receptor molecule on cell surfaces that recognizes the neurotransmitter adrenaline. But after he acquired one of the first ABI sequencing machines, he focused on a faster way of finding genes by generating expressed sequence tags or ESTs. This technique had been developed by Paul Schimmel and his colleagues at MIT to look for muscle genes as early as 1983. It involved extracting RNA from tissue and using it as a template to make complementary DNA, or cDNA, thereby isolating the protein-coding sequences of genes. Sequencing 150–400 bases at the ends of the cDNAs gives a pair of unique tags for each. Using these ESTs to probe the sequences already deposited in the public databases, you can find out if they represent known genes, if they have similarities with genes from other species, or if they come from previously unknown genes. By the spring of 1991 Craig had completed his first batch of ESTs derived from brain tissue, and he wrote to Jim Watson to propose that, instead of sequencing human chromosomal DNA, with all the difficulties of assembly and interpretation, the genome project might fund him to attack the protein-coding regions first by extending the EST approach. But the genome research center stuck to its policy of beginning by mapping out the whole genome. Although ESTs would later prove to be extremely useful in assisting that mapping effort, the study section that reviewed grant applications declined to support Craig's proposal.

Craig cannot have improved his chances by publicly positioning the EST strategy as a cheaper and more efficient alternative to genomic sequencing. He published his first paper on his EST work in *Science* in June 1991, while he was still at the NIH lab. In an accompanying news article, he described his approach as 'a bargain by comparison to the human genome project . . . We can do it for a few million dollars a year, instead of hundreds of millions.' Although he added that 'The cDNA approach does not eliminate the need for the Human Genome Project,' he was clearly trying to bring about a shift in policy through his public pronouncements—a tactic he has deployed frequently since.

We had already been personally affected by Craig's competitive streak. Some time that same year we heard that his lab was sequencing worm ESTs. I don't think I'm being paranoid in suggesting that they were trying to compete with us; there had been quite a lot of publicity about our getting the sequencing grant. At some point we held a conference call in which Craig argued forcefully that ESTs were a better way of finding genes, that he'd found a gene that we'd missed and so on. It's true that the program we used to hunt for genes automatically in the sequence was very elementary in those days, but I was in no doubt that in the long run sequencing the complete genome would be the only way to find all the genes. I saw Craig's challenge as a threat to what we were doing. If his lab was able to identify a substantial number of worm genes at a time when we had only a few dozen, because that's all there were in the clones we had sequenced, it would probably not help our case for future funding.

We had started thinking about sequencing our own ESTs in late 1990. Bob thought it would be a good idea to do a batch in parallel with the genomic sequencing; although it wasn't originally part of our program, we figured we could afford to do it. Chris Martin had just moved from the lab of Marty Chalfie (who was working with the mechanosensory mutants) at Columbia University to Bob's department at Washington University, and he had made a library of worm cDNAs. Bob persuaded him to let us sequence ESTs from this library, which he copied and sent to us so that we could do half

each, beginning in April 1991. We called our project a 'survey' of
C. elegans genes—it was not meant to be exhaustive, but to show the
range of genes. Our paper came out in early 1992, back-to-back with
a paper from Craig's lab on worm ESTs, in the first volume of
Nature Genetics. It was quite useful to have done it, but I also wanted
to put down a marker saying that we could do this just as well as
anybody else—and now we were going to get on with sequencing
the genome, thank you very much.

I saw cDNAs as a means to an end, not the end itself. And they
would never give you all the genes. In the news article accompanying his *Science* paper on human cDNAs, Craig estimated that he
could get 80 or 90 percent of the total number: in the same article, I
was quoted as saying 'Eight or nine percent is more like it.' (This
was a joke on my part—I just meant to emphasize that 80–90 percent was optimistic. I learned from that never to joke with reporters;
having a sense of humour is not part of their day job.) Equally
seriously, cDNAs alone give you no information about the
regulatory regions of the genome that turn genes on and off. For
complete understanding of the genome, there was only one option,
and that was to sequence the whole thing.

Of course, the National Center for Human Genome Research
could have regarded the EST work as complementary and funded
Craig as well as the other genome centers. But, perhaps unwisely in
the light of what happened later, they did not. Craig blamed genome
researchers outside the NIH for standing in his way. 'The extramural genome community did not want genome funding being used
on intramural programs,' he later told a congressional committee.
Frustrated at the lack of support within his own institution, Craig
looked for alternative sources of funds in the private sector. It was
almost exactly the same time, ironically, that Bob and I were in discussion with Bourke. We chose one side of the line, and Craig chose
the other.

Wallace Steinberg, chairman of the investment company

HealthCare Investment Corporation, funded Craig's new institute to the tune of $70 million. TIGR was to be a non-profit-making operation, and Craig would be able to publish his work and so stay within the academic research community. There was a quid pro quo: in parallel Steinberg set up a commercial company, Human Genome Sciences (HGS), to market the discoveries of TIGR, and appointed William Haseltine as its chief executive. Haseltine had worked on the sequence of the AIDS virus HIV at the Dana-Farber Cancer Institute at Harvard University, and had also become rich through involvement in a number of biotech start-up companies. The deal was that Human Genome Sciences should have exclusive access to TIGR's EST sequences for six months before publication, extendable to twelve months if sequences proved to be likely drug targets. Academic scientists would be able to look at the TIGR database freely after that, but the commercial company would have 'reach through' rights to any further commercial developments. Almost immediately the company sold an exclusive license for prior access to the information to the pharmaceutical giant SmithKline Beecham for $125 million. Craig, who had received shares in HGS from the outset, became a multi-millionaire almost overnight.

TIGR has remained a non-profit-making organization, and has been an important contributor to many publicly funded projects. Yet I saw the deal with Human Genome Sciences as compromising Craig's academic integrity. I felt he wanted to have it both ways: to achieve recognition and acclaim from his peers for his scientific work, but also to accommodate the needs of his business partners for secrecy, and to enjoy the resulting profits. This apparent determination to have his cake and eat it set a pattern for what was to follow, when Craig launched a privately funded effort to sequence the entire human genome (see chapter 5).

Craig's first priority at TIGR was to sequence human ESTs. Despite his frequent claims to have 'developed' the use of ESTs to find genes, he was by no means the first exponent of this strategy.

Sydney Brenner, who has never approved of large-scale genomic sequencing, had argued in favor of cDNA sequencing as an alternative to genomic sequencing during some of the earliest discussions about the HGP in the mid-1980s. (He used the excuse that 'we should leave something for our successors to do.') Sequencing cDNAs formed an important part of the work of his Molecular Genetics Unit in the late 1980s. What Craig did was to combine the EST strategy with the use of high-throughput methods based on the ABI fluorescence sequencers to speed up the discovery of candidate genes. This was no mean achievement, but was not a novel strategy.

TIGR did score a first, however, in sequencing the first free-living organism, the bacterium *Haemophilus influenzae* which causes chest and throat infections and meningitis in children. Craig was collaborating with Hamilton Smith of Johns Hopkins University, who had discovered restriction enzymes in the late 1960s and so launched the era of recombinant DNA technology. Smith and the TIGR scientists working on *H. influenzae* bypassed the mapping stage and shotgunned the whole 1.8 megabase genome at once, relying on a computer search for overlaps to carry out the assembly, following the approach Fred Sanger and his colleagues had used for viruses. The TIGR scientists went on to use this whole-genome shotgun method to sequence other pathogenic bacteria such as *Helicobacter pylori*, which causes stomach ulcers.

Craig's promotion of ESTs may not have found favor with the HGP, but it helped to launch a genome gold rush in the private sector. Suddenly the human genome looked like something you could sell. After HGS, one of the next on the bandwagon was the Palo Alto company Incyte Pharmaceuticals, later Incyte Genomics. After one of its co-founders, Randy Scott, read about Craig's EST work in the *New York Times*, he switched the struggling company's strategy from trying to isolate proteins that might make good drugs to producing a catalogue of human genes through EST sequencing.

Within a short time he had big drug companies queuing up to buy his product. Unlike Craig, Scott was a businessman first and foremost. The company would make its money first through selling access to the database, then from charging royalties on commercial developments from sequences that it had patented.

Five years before, genomic entrepreneurs had found it almost impossible to raise money. The Harvard scientist Walter Gilbert, who had shared the 1980 Nobel Prize with Fred Sanger for an alternative method of genome sequencing, had shocked his academic colleagues by trying to take the genome project private in 1987. But his idea of setting up a factory-like 'Genome Corporation' that would sell clones, data and sequencing services to academia and industry perished in the stock market crash of October that year. In complete contrast, from 1992 onwards genome scientists in universities found venture capitalists hammering on their doors. Many took the shilling. Gilbert bounced back by launching Myriad Genetics with Mark Skolnick of the University of Utah, dedicated to selling diagnostic and therapeutic products based on cancer-related genes. One of the genome project's biggest grantholders, Eric Lander at the Whitehead Institute at the Massachusetts Institute of Technology, with Daniel Cohen of Généthon in Paris, co-founded Millennium Pharmaceuticals in Cambridge, Massachusetts. There were many others.

There is much potential for conflicts of interest as a result of these agreements. The natural inclination of a commercial company is to retain exclusive control of its product, whether through patents or through commercial secrecy. The goal of the genome project, as I saw it, was to provide as much information as possible that could be freely used by everyone, public and private, to advance our understanding and develop new treatments. I didn't have a problem, and still don't, with companies protecting their rights to the inventions they sell—drugs or diagnostic kits, for example—but I thought there was a real danger if they were able to gain exclusive rights to the information contained in the sequence itself. That would mean that no one else further down the line would have any incentive to use that information in creative ways, and science and medicine would be poorer as a result.

From the earliest days of the Sanger Centre, we received visits from companies and entrepreneurs eager to make deals with us. My reply was always that we had nothing for sale, and for Bob it was the same. It worries me a bit that I may have missed opportunities for the U.K. by so doing. I was always aware that for Bob, as part of the vast capitalist empire that is the United States, this luxury could be afforded. There were plenty of other United States sequencing centers for the venture people to do deals with; but we were the only major one in the U.K., so what we did was decisive. The Wellcome Trust is a charity, so couldn't trade directly, but we could have sold licenses to our intellectual property indirectly through its technology transfer section. Indeed, I was under some pressure at the beginning, from Bridget and Michael, to consider doing so. This led to an exchange with Michael one noisy evening in the Blackford bar at Cold Spring Harbor (why is it that one has to travel 3,000 miles to have meaningful exchanges with one's colleagues?), when I said that if the data couldn't be released freely then I would resign. As it turned out, my view was shared by many of the governors of the Trust, and so it became policy.

In the worm community these issues had seldom arisen. No fortunes were riding on worm genes, so everyone (more or less) was happy that information should be shared. The map, and increasingly the sequence, had increased everyone's productivity and enlarged the field in a very effective way. But with the human genome it was a different story. The Sanger Centre began life in an environment in which commercial pressure was always going to be part of the picture. Those who were working to map particular human genes either expected to secure patents on them, or were terrified that someone else would beat them to it. It made for an atmosphere of mutual suspicion. With our declared aim of mapping and then sequencing some large fraction of the human genome, it was inevitable that we were seen in some quarters as a threat rather than a resource.

An example of this was at the chromosome 22 workshop in 1994, when the Sanger Centre group led by Ian Dunham was accused by another mapping group of withholding data, and of using markers that had been discovered by others without giving appropriate credit. In fact there was nothing underhand about Ian's approach— it was just that his perspective encompassed the whole chromosome rather than particular bits of it, and he had underestimated the desire of others to stake specific claims. The network of chromosome workshops had been set up by HUGO, primarily to collect and collate human genetic data; it was not easy for the participants to adapt to the new world of genomics. Personally, I saw the workshops as potentially valuable in integrating the two, but we had to get over the understandable resentment that we newfangled people were invading the geneticists' patch. Many of the groups involved would have liked to move forward into mapping and small-scale sequencing. But that would take too much time and cost too much. It was already clear that increasing the scale of sequencing was going to be the key to getting the costs down. I hoped that we would find a way for everyone to be involved somehow, but we had to face economic realities.

A key step in gaining the trust of the community was to make it an absolute rule that all the sequence produced at the Sanger Centre, whether from worm, yeast or human, would be immediately released into the public domain. This was exactly what we had done previously with the worm map. Anyone who came asking us to sequence a particular clone had to accept that we would not keep it back to give them time to look for patenting opportunities: as soon as it was finished, it would be deposited in the public sequence databases, and out there for everyone to see. By doing this, we would not only help science to progress; we would make the sequences in themselves unpatentable. Patenting opportunities would be open only to those who went on to do the real work of discovering what the sequences did and developing commercial products based on that understanding. Immediate release would also ensure that the sequence was straight away available to users, avoiding any wasteful duplication, and demonstrating that our own lab was not profiting from it—in the current jargon, it was being treated precompetitively. This had the practical benefit of both giving us immediate feedback on what we had done, and winning the support of the research community for a project that some had resented for taking so much money. In fact, sequencing projects account for less than 1 percent of the total biomedical research budget, and by making the data freely available we ensure that everyone can benefit from the results.

But all this, while truly a calculation that we made, makes the policy sound too purely pragmatic, or at least too short-term. The fact is that we'd come to realize that the genomic sequence we were producing and dealing with is more than a commodity. It is the essence of biological heritage, the instruction book for living things. We'd recognized it for the worm, spoken of it in our proposal for establishing the Sanger Centre, but now we had gone further. Now the information that was coming off our sequencing machines was the code of our own species. The only reasonable way of dealing with the human genome sequence is to say that it belongs to us all— it is the common heritage of humankind.

4 MEGALOMANIA

I WAS SHAMELESSLY UPSTAGING THE REST OF THE GROUP. FROM MY wheelchair I was trying to persuade the director of the Wellcome Trust, Bridget Ogilvie, as well as Michael Morgan and other Trust representatives, that we should accelerate our sequencing program immediately and cover most of the human genome by 2001, four years ahead of schedule. It was 1 December 1994. Five days previously I had been knocked off my motorbike and carted off to hospital unconscious, with a broken pelvis. Yet I was so carried away with excitement over this new idea that nothing was going to stop me—and it seemed that the Trust group was excited too.

It was Bob Waterston who had conceived the bold plan to accelerate the project. At the beginning of September 1994 he came over with his colleagues from Washington University for the annual joint lab meeting—'that fateful visit', he calls it. Bob found me rather worn out by politics and doubtful about how to go forward. As always, he stayed with Daphne and me, and we talked a lot.

It had been a confused year. The Sanger Centre was growing rapidly, and was becoming visible to the world of science. It was increasingly evident that we were not going to be a flash in the pan, and of course this meant that rivalries began to emerge, as in the case of chromosome 22. It was clear that if so much fuss could arise over just 1 percent of the genome, we were not going to get the job done if we carried on in this fashion. These were early days, though, and we still had a little time to sort out the politics. So it was well worthwhile continuing to try for general agreement.

Important as it was to get the human sequencing onto a workable footing, my own first priority was to get on with the worm sequence. Worm and yeast were the Sanger Centre's lead projects in terms of both our scientific credibility and technological development. But the worm work was funded wholly by the MRC. The Wellcome Trust, which at that time was putting up an imposing new building to be the Sanger Centre's permanent home, wanted to see more human sequence as soon as possible. Only human sequence would justify the large investment in such a high-profile project as the Wellcome Trust Genome Campus. However, it was too early: mapping was going forward, but space in the old building was limited, and there were as yet too many problems with finishing human sequence for that part of the operation to be easily accelerated. We were actually following quite closely the schedule in our original proposal, but there was pressure to do more. I'd given Bob some idea of my anxieties when we talked on the phone before he came over, but this was the first chance I'd had to discuss it face to face. He sat for a long time at the picnic table in my garden one late summer evening, listening sympathetically and obviously concerned that I seemed to be under so much stress.

Soon after Bob returned home to St. Louis, he sent me an e-mail headed 'an indecent proposal.' It contained nothing less than a strategy for completing the human genome in what by then current standards seemed an impossibly short timescale. It set out over several pages how we could each accelerate our production of sequence to 200 megabases per year (at the time we were doing ten or fewer). It went into great detail about the number of machines and people we would need, how often we would run the machines, and what it would all cost. It departed from our previous practice in proposing that we should churn out sequence as fast as possible and use automatic assembly and editing procedures to string it together to a reasonable but not absolute standard of accuracy—99.9 percent rather than 99.99. We would continue the slower process of hand finishing, but initially at little more than our current rate of about 10 megabases per year. In other words, he was talking about beginning with a working draft of the genome (but one much better than the one the world celebrated in 2000), and moving on to the gold standard, finished product only later. Bob reckoned we could get 200 megabases of almost finished sequence and 10 of finished sequence for a budget of $20 million per year, an average cost of around 10 cents per base—an order of magnitude less than the costs that were being discussed a decade before. Bob concluded, 'I would propose that you do 200 [megabases], we do 200 [per year] and we think about involving a third group and propose to do the entire genome between us in 5 years starting in 1996.'

It turned out that the idea had taken root in Bob's mind on the long, boring transatlantic plane flight back to St. Louis. He had a watch that had calculator buttons.

I sat down and started playing with the numbers—how many reads, how many clones, how many machines and so on. The hard part, we knew, was making the sequence perfect. I worked through these numbers and they weren't unreasonable, but what was impossible was the idea that it would be perfect. But I decided that for the human we could do enough, at least, to get things going. We already had a lot of experience with the worm of releasing unfinished sequence and finding that people were so grateful, and what a stimulus it was for research. I was also confident that our long-term commitment was to get it finished. So I got off the airplane very excited that we could do this, that it was plausible.

Rick Wilson bought it, though he commented sardonically that the St. Louis lab would stop Bob taking his calculator watch on airplanes in future: the combination of the buttons and an oxygen-starved Bob brain was clearly dangerous for everyone's sanity.

I too realized immediately that we could do it. In the Genome Sequencing Center at Washington University in St. Louis and the Sanger Centre at Hinxton, Bob and I headed the two most productive genome sequencing laboratories in the world. We each presided over an operation involving a couple of hundred people and dozens of machines, working a factory-like schedule to keep the sequence pouring out seven days a week, but at the same time constantly refining the tools needed to analyze it and improving the technology that generated it. By 1994 we were speeding up: at the Sanger Centre we were going to meet our target of 40 megabases of finished human sequence in the first five years, as well as finishing the worm and yeast genomes. Automation was improving all the time.

A crucial advance had come from St. Louis. Since 1989 Bob had been collaborating with a colleague in the Genetics Department, Phil Green. Phil originally trained in mathematics and was interested in taking a mathematical approach to solving problems in biology. When we started sequencing the worm, he immediately began to develop programs to help with assembly and gene finding. By the end of 1994, when we were thinking about scaling up our work on human DNA, he had designed a piece of software known as phrap (Phil's Rapid Assembly Program) that could assemble human sequence far better than previous assemblers. Assembly software for the human genome has to take note of the fact that half of it consists of bits of DNA that look similar, known as repeat sequences. Because of these repeats it is very easy to make errors in assembling the raw reads into a continuous sequence, but Phil's program makes sure that you get the best possible match for each repeat. In 1994 Phil had just left Washington University and gone to the University of

Washington in Seattle to join Maynard Olson; there the two of them lent their highly influential and independent voices to the debates over the strategy for the human genome that took place in succeeding years.

It wasn't just that we *could* get going on the human sequence; Bob and I both felt strongly that it *needed* to be done. Although the HGP had committed itself in 1990 to sequencing the whole human genome by 2005, it had begun with a raft of projects, including the worm sequencing effort, that did not produce human sequence at all. Less than 1 percent of the genome was represented in the public databases by the end of 1994, and most of that was in the form of cDNAs rather than genomic sequence.

Meanwhile the EST approach to finding genes was making rapid strides. For two years Craig Venter at TIGR and the Silicon Valley-based company Incyte Genomics had each been building up databases of these partial gene sequences. TIGR's arrangement with its associated company, Human Genome Sciences, meant that although academics could look at the ESTs in the TIGR database, HGS—or rather, its exclusive licensee SmithKline Beecham—had six months to a year of exclusive access to them first. Incyte's database was a proprietary product that you had to pay to use, and they also retained rights on further development. Although the NIH's ill-advised bid to patent the ESTs that Craig had discovered had been withdrawn, it looked as though private interests were finding other ways of locking up basic information about the human genome, moves which could only have a detrimental effect on future discovery.

In 1994 a white knight had come along in the form of the pharmaceutical company Merck, and in particular its vice-president Alan Williamson. The big companies weren't any happier than the academics that upstart genomics companies looked like cornering all the rights to valuable genome information. Merck funded a massive drive to generate ESTs and place them in the public databases, where they would be freely available to all. The effort involved a consortium of United States public labs, with the Genome Sequencing Center at St. Louis at its heart. Bob Waterston and Rick Wilson received a grant from Merck to generate 4,000 ESTs a week for two years, starting in January 1995. By doing this, Merck not only gave the entire research community, public and private, free access to valuable genomic data; it also made those sequences (and possibly the whole genes from which they came) much more difficult to patent. Once the sequences had been in the public domain for a year they could not be patented; and it would be tough for any company to identify the most promising genes out of so many and understand their function in such a short space of time.

I thought Merck's action was a great thing for science and a triumph for the principle of free access to genomic information. But Bob and I both knew that the EST database would not be enough to reveal all the genes, never mind to understand how the genome functions as a whole. The strategy depends on finding enough RNA to make the complementary DNA strand. Some genes are expressed in such small amounts, or such specialized tissues, that it is hard to find the RNA at all. But, much more important, RNAs correspond to only about 1.5 percent of the total DNA. The rest may not be directly involved in coding for proteins, but it is certainly not all junk. In order to understand how living tissues make use of genomic information in an integrated way, you have to do the whole thing. It had been hard enough to win the argument for whole genome sequencing back in the 1980s. Now, with the databases filling up with ESTs, there was a danger that people would question once again the value of sequencing the 'junk' between the genes. The best answer would be to do it, and to show how fantastically useful it was to scientists trying to understand the basis of disease.

Why wait? The HGP funders were hoping for some magic new technology to come along, and putting a lot of money into projects that promised to develop it. But meanwhile we were finding that the existing technology could do the job perfectly well. Running gels might seem fiddly and labor-intensive, but we were constantly finding better ways to use them. ABI was promising machines with more and more lanes—already we had modified their machine to run forty-eight lanes, and sixty or even ninety-six lanes were talked of. They and others were already beginning to experiment with capillary technology—running each sample through a fine, gel-filled tube instead of a lane in a slab of gel. There were people working on automating all the tedious and time-consuming jobs such as picking clones. We had excellent support from our bioinformatics teams in developing innovative software to track the samples and make sense of the results. It was all *working*.

Even so, I gulped a bit when I saw what Bob had written. But I quickly realized that it was not ridiculous. Bob is not a man given to wild impulses; on the contrary, he's a wonderfully stabilizing influence. His confidence was infectious. It immediately pulled me out of my gloom and set me thinking about how we could begin to set in train what I dubbed the MGP—the Megalomaniac Genome Project.

We knew from the start that one obstacle would be political. Although we were the biggest producers of sequence around, we were known as worm people. There were labs all over the world that had been mapping human chromosomes, more or less effectively, for years. They would undoubtedly expect that in the fullness of time they might move on to sequencing. Many already were, on a gene-by-gene basis, and one or two had funds to begin large-scale genomic sequencing. We could well be seen as interlopers, and resented accordingly. I saw this as being a particular problem in relation to the other international genome labs, such as those in France, Germany and Japan, and was anxious that any formal proposal should not be seen as too exclusive. Bob took the point, but put the alternative view that it would be impossible to accommodate the interests of everyone who thought they deserved to be involved. He pointed out that our bipartite collaboration on the worm was exceptional among genome sequencing projects in its success and lack of friction. And he admitted that the megalomaniac aspect of it was part of its attraction; sounding out colleagues such as Rick Wilson, he'd already found that the very audacity of the plan could be a powerful motivating force. But he agreed that trying to do the whole thing between the two centers would upset a lot of people, so we stuck with the proposal that we should do a third each, leaving a third for others.

We also knew that we had a practical problem. In the case of the worm we had a collection of mapped clones covering the entire genome, all ready to go into the sequencing pipeline. For the human, such a comprehensive set of suitable clones did not exist. There were technical problems in getting the DNA to clone stably in bacterial cosmids, as we had done for the worm. Although there were YAC maps in existence, they were of variable quality. The first whole-genome physical map of the human genome had been created by Daniel Cohen at Généthon in France. Cohen, who was director of the Centre d'étude du polymorphism humaine (CEPH) in Paris, launched Généthon in 1990 with funding from the French muscular dystrophy association as one of the first large-scale, automated genome 'factories' in the world. (The launch of Généthon had been one of the factors that predisposed me to the idea of a development such as the Sanger Centre in the U.K.) Cohen had moved fast to produce YAC maps, first of chromosome 21 and a year later of the whole genome. At the same time, his colleague Jean Weissenbach had produced a whole-genome genetic map, three years ahead of the timetable set by the HGP. But while the genetic map has remained one of the main anchors of the project all the way through, the physical map turned out not to be such a useful resource.

The problem was directly related to the nature of human DNA. We had been able to use YACs for the worm map, once we had sorted out problems of contamination with yeast DNA and learned to deal with the large size of the inserts. But the number of repeats in human DNA—as much as 50 percent of the total is in repeat sequences, compared with 15 percent in the worm—caused constant problems in getting stable clones. They would break up and rearrange themselves as the yeast cells reproduced. The Généthon map had particular problems because Cohen had developed a method of making 'megaYACs' up to 1 million bases long, which only increased the opportunities for chimaeras and deletions. New bacterial cloning methods had recently been developed, such as bacterial artificial chromosomes or BACs, which held smaller inserts and looked as if they would clone more stably. But there was a lot of work to do to produce a library of BACs and map them before they were ready to go into the sequencing machines. It was something we thought we'd be able to crack if we put our minds to it, although we were admittedly a bit vague on exactly how to go about it. We reasoned that there was no need to complete the map first, as we had for the worm; we could map and sequence in parallel.

Obviously, we could not launch such a huge effort on the basis of the funding we had. I was fairly optimistic that we at the Sanger Centre would be able to get what we needed, as the Wellcome Trust had gone into genome sequencing with the clear intention of doing substantial amounts of human sequence. I hoped it would just be a question of persuading them to bring some of the money forward. For Bob it would be more difficult. He had already had to fight a battle with the NIH, which had upped his funding for the worm in 1993, to be allowed to spend some of his grant on small amounts of human sequence, in parallel with what we were doing—the study section that reviewed his proposal did not believe that the methods would work on human DNA. 'In the end they more or less threw up their hands and said they didn't think we should do human, but it would be up to us to decide,' says Bob. 'And so we did a little bit of human—but we knew we were being watched.'

Because of the political sensitivities we didn't want to make a big splash with our idea until we'd done a bit of informal sounding-out. Among others Bob talked to Francis Collins, who had by then been head of the NIH genome program for a year, and Maynard Olson, who could be relied on to spot any flaws in the plan. 'Nobody could shoot it down,' he says, 'although Francis was very cautious.' They all seized on the problem of clone supply, which of course was our weakest point. On the political question, there was a widespread view that no one else was yet in a position to take on the remaining third of the task. Meanwhile I talked to Jane, Richard and David, since they would have to bear the brunt of managing a ramp-up in sequencing, data management and mapping. They were immediately enthusiastic, and in a very short time Bob and I had arranged to make presentations to the Wellcome Trust and the Medical Research Council in the U.K., and to the National Institutes of Health in the United States. Bob was going to come over and back me up at the Wellcome and MRC meetings.

On the Saturday before the date set for the meetings in London, I set off from home for the lab on my motorcycle at about seven-thirty in the morning. I was still a member of one of the worm sequencing groups, and took my turn loading machines at weekends. I had ridden a motorbike as a research student, and had recently taken it up again; the LMB was only a short cycle ride away from Stapleford, but Hinxton was further from home, and I also liked the fact that I could get to the airport on the motorbike more quickly than I could by car. That morning, at a junction a mile into my journey, a van ran into me from the side. I remember nothing of the accident, but evidently the impact pushed my leg back through the pelvis, breaking both the pelvis and a ligament in my knee. My 550cc bike was left as little more than nuts and bolts, and I was flung right across the road. But I was incredibly lucky. A woman in the car behind phoned for an ambulance and I was taken straight into the operating theatre at Addenbrooke's Hospital in Cambridge. A&E was empty at that time in the morning, and the duty surgeon was Dennis Edwards, one of the best in the business. The nerves were intact, the injured tissue had had no time to die, and he reset the pelvis in perfect alignment, holding it all together with a plate and screws, which I still have.

When I came round, still groggy from the anaesthetic, I tried to get out of bed and go and load the sequencing machines. Then when I realized I couldn't move, I thought I had to try and find someone else to go and do it. That may have been a little irrational, given what had just happened. But as my mind cleared, I was gripped by the thought that I had to get to the meeting at the Trust on Thursday. As soon as I was conscious Bob came on the phone, trying to get me to give some sort of account of how badly hurt I was. I just kept saying, 'Oh, it's OK, this can't stop us, there's no way we can change our plans now—when are you coming over?' He agreed that he would come, although he says now his main anxiety was to find out how I really was. I spent the next three days battling with the doctors to let me leave hospital in time for the meeting. In fact I made such a nuisance of myself in the ward, constantly using the phone, holding meetings with colleagues and so on, that they put me in a private room. They said I couldn't leave until I could get around on crutches, so I practiced doggedly until they reluctantly agreed to let me go. On the morning of the meeting I got myself up in the pitch dark, dressed and maneuvered myself into my wheelchair with my crutches, and went down in the lift. Bob was staying with Daphne, and David Bentley was there overnight too, ready to go down to the meeting the next day. They had arranged a car and a driver to take us to London, and the hospital discharged me into Bob's care. I told them he was a doctor, which was true, although he hadn't practiced since finishing his training.

And so I arrived on time at the Wellcome Trust meeting—thanks to Bob, David and the wheelchair. I was so hyped up I felt I could have sold anything, and the meeting seemed to go very well—well enough, anyway, that they invited us to submit a formal proposal. We left in high spirits, Bob pushing me down the Euston Road to the MRC, stopping for a pub lunch and a quick turn round the rose garden in Regent's Park on the way. Bob thought the afternoon's meeting was less successful than the morning's. But the MRC secretary Dai Rees gave us a fair hearing, and I certainly felt that we were being taken seriously. All in all it seemed like a good day's work.

The surgeon, Dennis Edwards, had promised to have me back on the bike within six months; I was more concerned about being able to go hill walking. Ever since I was a student I've got out for a few days' walking each year, often in the Lake District or Scotland, and lately my son Adrian and I had taken to going together for a long weekend in the spring, scrambling around the crags of the west coast. But I need not have worried. Just the following summer Daphne and I stood on Mount Brandon, 3,000 feet up, looking out over the Atlantic in south-west Ireland; I sensed that Edwards' handiwork was perfect, and that he'd given me back the beauty and solitude of the high places. Adrian now lives in Edinburgh, where he works in software, and from there we continue to have wonderful excursions.

I didn't replace the bike. Bridget Ogilvie and her colleagues at the Wellcome Trust were horrified when they discovered I'd been using it to get to the airport—'After we'd invested all that money in this bloke!' she says—and forbade me to do it any more. More significant from my point of view was that Daphne, too, felt that this had gone far enough. But I didn't need much persuading. It makes life pretty important that you've survived something like that. I don't mean that I had any religious sense of having been preserved for a purpose. But I certainly felt that I was very lucky—not only to be alive, but to have been put back together again as good as new—and that I had better make the most of it. I was back at the lab within a week, and turned in my crutches three months later. In the interim, I shared a taxi into work with Matt Jones, one of the group leaders at the Sanger Centre, who had been knocked over not long after me and was recovering from a fractured femur.

Two weeks after our London meetings, Bob presented our idea to a gathering of genome lab heads at Reston in Virginia, which had already been scheduled to discuss future strategy on the human genome. Most of them had no inkling of what we had planned; we'd taken only a few into our confidence. Bob was down to speak, and naturally they assumed he was going to talk about progress on the worm genome. Instead he produced a set of handwritten overheads he'd made the night before and laid out our proposal. 'They were taken aback,' says Bob. But Maynard Olson stood up and said that what was on the table was a realistic plan to sequence the human genome, the first anyone had produced, and that they should take it seriously. The debate that followed focused on two issues: the map, which Bob knew was a problem ('I didn't quite say so, but I thought that was their responsibility!' he says) and the value of having a draft—an intermediate product on the way to fully finished sequence. Yes, the worm community had confirmed that unfinished sequence was useful. But there was a real worry that once the draft was available everyone would lose interest in finishing, and we would never end up with an archival product that would stand the test of time. On the other hand, Francis Collins had seen that the plan provided a lever with which he could increase his budget, and he had arranged for Bob to talk to Harold Varmus, the NIH director, just before the Reston meeting. Francis knew that sooner or later there would have to be more money for production of human sequence. The wide distribution of smallish grants for mapping, technology development and small-scale sequencing projects that he oversaw at the time would not meet the need. Even giving everything in the pot to a smaller number of people would not solve the problem; the pot itself needed to be bigger.

Now that the plan was out in the open, everyone was talking about it. The important point was that no one seriously challenged Bob's math. His timescale for getting a 99.9 percent accurate sequence seemed credible. But in addition to the mapping and finishing anxieties, political worries began to circulate, and these were even worse than we had expected. If the funding agencies put so much money into a few big centers—not necessarily the same ones that were doing the mapping—what would happen to the rest of the twenty or so labs in the United States (and a similar number in the rest of the world) that got left out in the cold? In an interview about our plan published in *Science* magazine in February 1995, I argued that, given that money was tight, we had to go for the most cost-effective option. 'The entire bill for biomedical research would be lower in ten years' time if we start now than if we delay,' I said. 'If we don't start now, there will be innumerable other gene hunts and sequencing projects going forward, taking collectively an enormous amount of money.' If it can be done now, I argued, 'Why fiddle around?' Our thinking at the time was that as the draft was produced, anyone who had an interest in a particular region could get the relevant clones from us and finish them to a higher standard.

The debate continued at the May 1995 Cold Spring Harbor genome meeting. The organizers (who included Bob Waterston and David Bentley) decided to break with tradition by having not one but two 'keynote speakers' to round off the meeting just before the farewell banquet on the Saturday evening. They invited Maynard Olson and me, Maynard to speak about mapping and me about sequencing. Maynard spoke first. I had been a bit nervous about what he would say, because despite his support for Bob at Reston, he had certainly been very doubtful about the idea at the beginning of the year. Always a perfectionist, he had been afraid that producing the draft would be a distraction from the real goal of fully finished sequence; and he thought that if money were taken away from developing new technologies to put it into production sequencing we would never reap the benefits of a fully automated process. But by May he'd had a chance to talk it over with Bob, and was 'generally satisfied' that our plan would not undermine the long-term goal. Maynard and I shared a house on the Cold Spring Harbor campus during the meeting, and had a chance to coordinate our messages. We sat at the same table preparing our talks, with me scribbling out my usual handwritten overheads, he going off to the conference office to get his properly printed.

Maynard backed us in his talk, though still expressing reservations about the quality of the product. I was then able to get up, lay out our plan to produce what we were calling a 'sequence map' by 2002, and say: 'Let's just do it.' (To Maynard's mortification, his beautifully produced slides were printed right to the edge of the sheets and did not fit on the screen. I could not resist using the incident to contrast our styles of work. 'My slides may not be as smart as Maynard's,' I said, 'but at least they fit on the screen!')

Not everyone was convinced. For instance, Eric Lander, director of the Center for Genome Research at the Whitehead Institute, was standing next to me in the lunch queue one day at the meeting, and told me that he thought it was too soon to launch a bid for the whole genome. Eric is a big, ebullient New Yorker with ambitions to match. He had started out as a mathematician and taught for a while at Harvard Business School, but in the 1980s he had switched to medical genetics and joined the Whitehead Institute. He became convinced that the quickest way for the field to move forward would be to make a big investment in infrastructure, in the form of genetic and physical maps. He worked with Helen Donis Keller and Phil Green, who were then at the private company Collaborative Research Inc., in the first attempt at a genetic map of the human genome, placing about 400 markers in total. In 1990 he received funding in the first round of HGP grants to establish a Genome Center at the Whitehead and to make first mouse and later more human genomic maps. The center had established a high profile with an approach that depended on large-scale automation. In 1995 he began to turn his attention to high-throughput sequencing, the obvious next step.

> We came at it from the point of view that we hadn't sequenced a damn thing but it didn't matter. What we did know how to do is to take on large problems and scale them up and get them done. Our argument to NIH was that although there are groups in the United States that have been doing sequencing already, they're doing it as a cottage industry, and you want at least one group that will take it on in a scalable fashion. We were always trying to walk the fine line of taking on projects that are a little too audacious but not so audacious that we get laughed at. So we began automating the hell out of sequencing.

At the time of our proposal, Eric had only just begun to plan for sequencing and would need more time to develop the same kind of highly automated approach to sequencing that he had taken to mapping. At the same time he was critical of our proposal because we wanted to continue on our course, automating as we went along. Through our success so far we had demonstrated that this approach worked. Each year our production had gone up and our costs had gone down, and naively I assumed that our results would speak for themselves. But Eric's doubts would carry weight—much more weight than I realized at the time—and he was not alone in his views. Today he still argues that our bid was premature.

> They were out of their minds, because they weren't automated . . . My objection to scale-up then was that it wasn't a serious scale-up. I thought we needed to develop a system that could scale up twenty-fold rapidly, and even more eventually. I felt we should wait a couple of years and get systems in place that would let us scale up . . . Nonetheless, what Bob and John did was very valuable, because they

began to force the question, and even if it wasn't planned to start sequencing, all it needed was a couple of years to mature the thing and that's what happened.

Eric may have needed another two years, but we didn't; we were ready to go ahead as we were. And, despite his reservations, Eric began to talk to us about joining our collaboration—in effect, becoming the third partner that our proposal envisaged. In July we had a formal exchange of letters, agreeing to share expertise and to hold small group meetings of people from the three labs who were working on the same problems.

Eric is the first to admit that ambition is one of his defining characteristics.

Back in 1995 I got myself in a lot of trouble by saying the genome project was going to get done by two or three centers. It was a problem of scale. You could not hit an efficient scale in twenty places—it was dotty to imagine we could have twenty centers inventing everything. My orders to my troops were that we should be prepared to sequence the entire human genome ourselves—not that we should do it, but that we weren't going to start the project unless we were building a system that—if called upon—could.

Meanwhile we knew what we could do, and didn't think Eric's criticisms were justified. We pressed on with preparing our proposal. Jane, Richard, David and the rest of the BoM put together a massive document outlining a program of work for the Sanger Centre from 1996 to 2002, including sequencing one-third of the human genome. Of course, this included our own plans for automation which would go on alongside the scale-up. I added an introduction.

Our thesis is that the time has come to make a major assault on the human genome. There is no need for further hesitation. All that is needed is a common will, internationally rooted, for the sequence map of the human genome to be achieved by the year 2002.

I sent the proposal, asking for an additional £147.2 million over seven years, to both the MRC and the Wellcome Trust. This was for fully finished sequence. Everyone agreed that the MGP was only a step on the way, so it made sense to budget for the finished product. Asked by the MRC what would be the consequences if we were funded to do one-sixth of the genome, only half as much as we had projected, I replied, 'In such a climate, the Sanger Centre's influence in driving forward the international program would be lost. As a result of this it would be unlikely that the major uncharted regions of the human genome would be completed systematically and in the public domain in the foreseeable future.'

The proposal came up for consideration by the Molecular and Cellular Medicine Board of the MRC, and the Scientific Committee of the Governors of the Wellcome Trust in the autumn of 1995, a year after Bob and I had first begun to put our plan together. At the end of September a joint MRC/Wellcome committee descended on the Sanger Centre to review our work and collect information for a report to the two funding bodies. Mike Dexter, who was then deputy director of the Paterson Institute of Cancer Research in Manchester and chairman of the MRC's Molecular and Cellular Medicine Board, chaired the delegation. It included, as an 'observer' for the NIH, Eric Lander. I was somewhat surprised when I heard he was to take this role, given that he had already signed an agreement to collaborate with us.

It was a strange occasion. We had invited Bob to come over, to emphasize our international collaboration, just as he had invited me for his NIH review board a few weeks previously. There I had sat at the back most of the time, but I had been called to get up and say a few words about how much we valued the link and what we had been able to achieve. But at our review Bob was not allowed to

speak. The committee sat behind a long table in the rather nice conference room that we had inherited from Tube Investments at
Hinxton. At that time there was a vogue among accountants for
large chunky calculators with big square keys. I always thought on
seeing these things that the message was: 'I have big square buttons
and I deliver big square numbers that are better than your little
round numbers.' Murray had one, in keeping with his position, and
so did Eric on this occasion. He used it repeatedly to challenge our
calculations, casting doubt on our ability to scale up the production
of sequence on the timescale we envisaged and questioning our cost
estimates.

I was confident that the figures Jane had prepared for the proposal, based on her considerable experience of running a sequencing
operation, were as good as they could be. It turned out later that her
predictions of future costs were accurate too. But I found I couldn't
stand up to Eric, at least not in public and not with his calculator. In
addition to his intense inquisition of me, I remember being quite
open about the uncertainties. I said things like, 'I don't exactly know
how much this is going to cost, but let's put an outer envelope on it
and see how it goes.' Truthful it was; but it wasn't the official way to
play the game. Bob put it down to my lack of experience with this
kind of review, which was much more like the way they did things
in the United States. You really needed to have all the facts and costs
at your fingertips, or at least look as if you did. Up till then I'd happily operated with the usual British approach to funding, which Bob
characterizes as: 'If you think I'm a top-notch person and this is a
high-class project then you should fund it. As far as the details go,
you just have to trust me.' Bob knew what was needed, but he wasn't
allowed to speak, and must have despaired as he listened to me. But
in the event Eric's interventions and my unfocused responses probably had less influence on the outcome than I imagined.

Later the same day Mike Dexter came to see me and said that the
committee was supportive in principle of our doing a third of

the genome, although they were not going to recommend that we be funded to the tune of the full amount that we had asked for. By now the two Sanger Centre backers, the Wellcome Trust and the MRC, were even more unequal partners than they were at the beginning. Our understanding all along was that they would share the costs of the program equally, and the MRC had certainly been participating in the discussions of our future strategy as an equal partner, up to the day of the meeting. What happened was that the Wellcome Trust gave us the money to produce fully finished sequence of one-sixth of the genome by 2002—a half share, as they had promised, about £60 million in total. The MRC, on the other hand, gave us just £10 million over five years, some of which was to finish the worm.

With hindsight, I suppose it was unrealistic to expect that the MRC would come in with equal funds. Almost all of its budget was spoken for, in long-term funding to research units and programs, before it could even think about giving out money for new projects. Diana Dunstan, director of research management at the MRC, confirms this view.

> It was a question of finance, not necessarily the immediate needs but the longer-term implications and the probable impact of MRC involvement on the funding available for our broad portfolio. In other words, if we had made a commitment to fund [the proposal] then there would almost certainly have been very little left over to do other things. With hindsight, this was probably a correct analysis.

And getting a substantial budget increase from the government was not easy. In fact, following my December 1994 meeting with Dai Rees, the MRC had remarkably quickly won additional funds from the government's Office of Science and Technology, specifically to support 'the U.K. arm of the global initiative to generate a sequence map of the human genome in five years.' It was ironic that the U.K.

government became the first body to put money on the Megalomaniac Genome Project in its original form, although the sum was far short of what was needed. The grant amounted to £2 million per year for five years: in other words, the MRC put nothing more from its own resources into large-scale human sequencing at the Sanger Centre. In addition to the existing worm grant and the new government money (the first two years' installments for the worm project, all the rest for human sequencing), we had a couple of other MRC grants to David Bentley for human genetics projects, but that was it. The MRC had started the whole field of genomics in the U.K. and had seen the worm sequencing through, but from this point on it became very much a minor player in the Sanger Centre and before long it ceased to be represented on the board of directors of our management company, Genome Research Limited.

Dai Rees, who was then secretary of the MRC, puts the collapse of the partnership down to the inevitable constraints under which a government agency operates.

> The ideas of Wellcome on the scale, speed and style of development mushroomed beyond anything that could be realistic in terms of the public purse ... Our difficulties were not merely with funding the science, but with the investment required for the ambitious and imaginative development of the Hinxton site. Eventually I think the issue was forced by the irreconcilability of the urgency and freedom on the Wellcome side with our need to proceed within the frame-work and timescale of the processes of central government, in which MRC claims had to be balanced against other Research Councils, at a time when the philosophy of public spending was much more rigid than it is now. Also, I think we gave the appearance of hanging back because we didn't manage to get it across to the Wellcome that whereas they could make definite commitments, we could not do so ahead of Government decisions and could only express informal intentions.

The Trust did not feel it could make up the shortfall and fund the full third of the genome on its own. Today Mike Dexter, now the director of the Wellcome Trust, supports that decision.

In the mid-1990s, the costs were pretty horrendous. As time went on, the costs came down and the efficiency of the machines got better. We could have poured money in then, and got old-fashioned machines and expensive sequencing reagents, but it would have been partly a waste of money. By waiting two or three years, we could do it better and cheaper. Many people felt, what's the rush?

This, of course, was my reason for being vague at the time. The big square buttons and the NIH-style review were meaningless—or, at best, they were a sort of theoretical examination. Nobody could know what the picture would be like three years later, though in fact the guesses that we put forward that day were about right. The initial annual spend would not be all that high because it would be a matter of mapping and starting to ramp up the sequencing, not of jumping straight to full production. By the time of full production in 1998 there might well be better and cheaper technology; but instead of coming to it cold as actually happened, we would be poised and efficient. The best way would be to proceed optimistic-ally and openly, justifying each year as we went along. As it turned out this is what happened in practice, but at the time I failed to put the point across.

Michael Morgan, who unlike Mike Dexter was a Wellcome Trust insider at the time of the review, remembers it differently.

We [the Wellcome Trust] wanted to take a risk, we wanted to be adventurous. Costs were not the issue—the question was, could it be done or not? There was nothing in the review to make the MRC think they should not have been doing 50 percent. But we could not make up the shortfall—the £60 million would already be a very significant slice of the Trust budget. It was very brave of the governors to support it.

At the time this was the biggest grant they had ever awarded, and more than any sequencing center in the world had been promised. In the absence of any other initiative it sounded very good to say that the Sanger Centre would sequence one-sixth of the entire human genome (at a time when human sequencing had still barely begun). And I suppose I should have seen it as a vote of confidence (although if we had got any less than we did, it would have made a nonsense of the Trust investing so much in the Hinxton site). But to me, and to everyone at the Sanger Centre, winning only half of what we had asked for was a serious setback. It was not enough to drive progress as Bob and I had envisaged.

I remembered the advice John Smith, a pioneer of nucleic acid analysis at the LMB, had given to Alan and me one night in the Frank Lee when we told him that we were going to sequence the worm genome by the end of the decade. 'That's too long,' he said. 'You can't make a science project last that long.' The moral is that if you find yourself in possession of technology that allows you to go at a certain speed and you go at less than the maximum, then you're done for, because somebody else will come and do it ahead of you. I had made a similar point when the MRC asked what would happen if we got only half the money, and again at the review. But it hadn't worked. Between us in that review room we had missed our chance, and a couple of years later we were to face the consequences.

Meanwhile in the United States Bob Waterston was having no more luck. It seemed that the national funding bodies, the MRC and the NIH, were of one mind in erring on the side of caution. Bob's bid for funding to sequence a further third of the human genome was given a doubtful review. The committee had no quarrel with the quality of what Bob was doing; but it argued that it would be a mistake to spend a lot of money immediately on the existing approach when in three years' time there might be a better way of doing it. Bob gave a spirited reply, saying that it would be disastrous to delay for three years, but to no avail. He was awarded a quarter of what he had requested in the first year, with modest increases to follow.

Bob shrugged off his disappointment, believing that his detractors were simply postponing the inevitable, as he wrote to me in December.

> They don't want to see their opportunities cut off and can't admit that the human genome is going to be sequenced in the next few years, regardless of what the [National Center for Human Genome Research] does. I am really beginning to believe this. With the improvements that the two labs are making in software and methods, an ABI machine becomes an ever more potent weapon . . . Anyway I may be exaggerating some but major inroads will be made in the next 3–4 years, I'm convinced.

As in the case of the Sanger Centre, on the face of it he had little to complain about. When Francis Collins announced funding for a series of pilot projects to 'explore the feasibility of large-scale sequencing of human DNA' in April 1996, Bob received a third of the $20 million or so awarded for the first year. Five other labs were funded, including Eric Lander's and Maynard Olson's. The pilot projects were charged with producing a modest 3 percent of finished human sequence in the first two years; they were to be judged not so much on output as on what they could achieve in terms of accuracy, speed and cost-effectiveness. After a review of these projects in 1998, the funding agency would then decide on a strategy for the final phase.

It was all very admirable in its intention that the ultimate goal should be a complete sequence finished to the highest standards. But it was deeply frustrating to me that the biological community should have to wait until 2005 for this prize. It wasn't that I didn't think the quality of the final product mattered. But I did believe, very strongly, that making a less perfect product available sooner would be better for biology. People could get on with finding genes and understanding how they worked as soon as they had a draft, as we knew from our experience with unfinished worm sequence. And I still think Bob and I were right. There was no miracle new technology. Instead, the tried and tested gel-based approach simply became more and more efficient, costs falling from £0.50 ($0.75) per base to below £0.10 ($0.15) per base in five years, while output from the main centers increased twentyfold. In the light of what happened afterwards—the launch of a private-sector competitor for the Human Genome Project and the subsequent adoption of a 'working draft' strategy by the HGP (see chapter 5)—I have no doubt at all that the funding bodies made a big mistake in not supporting us at the time.

Francis Collins thinks the cautious approach was justified, on the grounds that we needed 'to be sure we knew how to do it at high quality before trying something intermediate.'

> Some people looking back in history might say there was an over-emphasis on this compulsive attitude about having every I dotted and T crossed. I actually have no regrets about the way we did it. I think we learned a prodigious amount by pursuing the pilot phase and focusing on high quality finished data . . . We had to bring on other centers [in addition to the Sanger Centre and Washington University], and some of them had not had the same experience, so emphasizing quality up front was a good way to be sure we didn't end up with a mess.

As *Science* reported at the time, the priority of the National Center for Human Genome Research seemed to be to 'hedge its bets and circulate its limited funds to a variety of labs.' Some outcomes were positive:

Eric Lander succeeded in getting his highly automated shotgun production lines running, and this later paid off for him in generating draft sequence quickly. But time was lost in getting on with the job: the public side was seen to be slow, and this subsequently made for a weak PR position. 'It would have been much better to have accepted John and Bob's idea at the time,' says Jim Watson now.

Even though we were not able to expand our sequencing effort as fast as we wanted, human sequencing really began to take off in 1995–6. As Bob and I had been hatching our megalomaniac plans, we looked at what was happening around the world and recognized that it was pretty chaotic. People were duplicating what others were doing; everybody was piling into regions of the genome reputed to have genes for cancer or other major diseases; there was no consensus on whether or how the sequence data should be made available to the community. If your goal is a complete genome sequence you can either do it yourself and ignore the rest, which means you have to go faster than everyone else, and you have to succeed; or you have to get everybody together. To a certain extent Bob favored the first option; he really thought we could split the whole thing half and half and do it all. But I've always felt that if you can share you should do so—that way you don't have silly competitions.

I also felt that there was a need to strike a blow against the gold-rush mentality that was developing towards the genome. There was a sense, largely but not exclusively fostered by the new breed of genome-based private companies, that everyone was in a race to stake claims as fast as they could and reap huge profits from their discoveries. A case in point was the public-versus-private drama that was beginning to unfold at about the same time over the breast cancer genes, a drama in which the Sanger Centre was playing a supporting role.

Mike Stratton, who now runs the Cancer Genome group at the

Sanger Centre, was then leading a team at the Institute of Cancer Research in Sutton, Surrey, dedicated to finding genes that place women at high risk of developing breast cancer. One gene, BRCA1, had been located by Mary-Claire King in the United States. But it could not account for all families with clearly inherited susceptibility to the disease, so Mike and his colleagues set out to find another. In the summer of 1994 they located the gene, which became known as BRCA2, on chromosome 13. Up to this point Mike had been collaborating with Mark Skolnick at the University of Utah, who was cross-checking the massive genealogical database compiled by the Mormons against cancer registries in order to identify families that could help in tracking cancer genes. Skolnick had set up a private company called Myriad Genetics (see p. 110) specifically to go looking for cancer genes and to market genetic tests for any that they found. Not long before his own group located BRCA2, Mike went out to talk to Skolnick to find out exactly what Myriad's role would be if the collaboration did indeed find and clone the BRCA2 gene. The answer was that Myriad would patent it and own exclusive rights to exploit it both for diagnosis and therapy. Mike was very much opposed to this idea.

> I was concerned about what would happen if in the future there was a conflict between the clinical or ethical imperatives and the commercial imperatives. Myriad has a duty to service the needs of their investors. I realized that I would have no influence on how the discovery was used. So after we published the location of the gene, I ended the collaboration.

Mike and his colleagues were now racing the Utah lab to find the gene itself and clone it. At that point the Sanger Centre became involved, when they came to ask David Bentley if he would make a clone map covering the approximately 1 megabase region where they knew the gene lay. He was happy to do that. But, having got the clones, David realized that the BRCA2 region would make a nice pilot project for the human sequencing program at the Sanger Centre and Washington University. He asked Mike what he thought. The sequence would be very valuable; but at the same time, the Sanger/Washington University data release policy meant that as soon as it was complete it would be publicly released, and that might help Mike's competitors. David was very clear that we would not sequence the region if the Institute of Cancer Research team were unhappy about it. 'We discussed it,' says Mike 'and decided that as we were in this game to get the gene found, it would be ridiculous for us to even consider not allowing David to go ahead.'

David was soon able to give Mike a date, November 23, 1995, when the sequence would be placed in the public databases. Shortly before that, the ICR team found a mutation from one of their breast cancer families that looked as though it might very well sit in the BRCA2 gene. Within two weeks of the sequence being available, they not only confirmed this mutation but found five more. There was now no doubt: they had found the gene. Mike moved fast to publish the group's discovery in *Nature*, while keeping it secret even from his collaborators until the last possible minute. But despite his efforts, enough information about the discovery reached Skolnick to enable him to locate the gene himself and bang in a patent application—the day before the ICR paper came out in *Nature*.

Despite his reservations about patenting, Mike had realized that in such a competitive area, the only way to protect his team's discovery from commercial exploitation by others was to patent it himself. The Institute took out one patent on the first mutation as soon as it was discovered, and another later covering more mutations. Meanwhile Myriad's patent applications claimed rights to the whole gene. The Utah scientists had been the first to clone BRCA1, on which they also own patents. They set up a commercial diagnostics center in Utah and, once the patents were granted, threatened legal challenges to any lab elsewhere in the United States that was using either gene to carry out breast cancer screening. All such screens henceforth had to be done at their own center, at a cost of around $2,500 per patient. Other labs could apply for licenses to carry out simpler tests to look for single mutations, but again had to pay a fee of a couple of hundred dollars per test. One of these tests is for a mutation in BRCA2 found by Mike's team, which is particularly common in the Ashkenazi Jewish population (those who originated from Central and Eastern Europe). 'The Ashkenazi BRCA2 mutation was in our original paper,' says Mike, 'so Myriad is claiming a fee from all women who undergo tests in the United States for a mutation that was discovered by us.' As an Ashkenazi Jew himself, Mike found this particularly hard to take.

In Mike's view it is 'unfair and unethical' that Myriad should have complete control of screening for breast cancer susceptibility, especially when (as Myriad acknowledged in its 1996 paper) their findings on BRCA2 depended on the earlier work of Mike's team. Only the Institute of Cancer Research patents stand in their way, and in fact Myriad has been less successful in Europe than it was in the United States at challenging the use of the gene. But having accepted massive financial investments, the company now has no alternative but to market its goods as aggressively as possible. And as a body largely funded by coins dropped in the tins rattled by Cancer Research Campaign volunteers, the ICR cannot justify spending the huge sums on lawyers that would be needed to fight Myriad through the courts.

I see this as a cautionary tale. In my opinion commercial imperatives have overridden clinical and ethical imperatives, just as Mike feared. By claiming rights to diagnostic tests for the two BRCA genes and charging for them, Myriad adds to the total bill for health care. In the United States the tab gets picked up by the patient, through higher insurance premiums; in the U.K. it's the taxpayer who foots the bill through increased costs in the National Health

Service. But much worse than this is the impact on science and future treatments. Once scientists really understand how the mutations in BRCA1 and 2 unleash the growth of tumors, they might be able to devise new therapies. But because of its patents, only Myriad has the right to market such therapies. Others with relevant expertise will have less incentive to join the effort; even if they make new discoveries there will be a lot of work for lawyers involved in sorting out cross-licensing agreements. The total amount of brainpower focused on the problem will be less as a result. It seems to me that companies such as Myriad are going for short-term profits at the expense of long-term benefits to human health—the promised benefits that are the ultimate justification for the whole genome sequencing enterprise.

Although in 1995–6 the full consequences of Myriad's aggressive approach had yet to emerge, it was pretty obvious where a focus on commercial profit would lead. Bob Waterston and I felt there was a need to get some kind of commitment from the international sequencing community that genomic information would be made publicly available and not parcelled out among the claims of the profit-mongers, with individual deals between companies and researchers. Michael Morgan, who in 1994 had been deeply involved in discussions in HUGO and elsewhere about the problem of open access to ESTs, was confident that the Wellcome Trust could play a role in bringing this about. He, Bob and I came up with the idea that what was needed was an international meeting to hammer out a strategy for deciding who would do what, and how the data would be managed. Michael got Bridget's agreement that the Trust should sponsor the meeting. In discussion with the other main funding bodies, the National Institutes of Health and the United States Department of Energy, he settled on Bermuda as a suitably neutral location (British territory but within easy reach of the United States), and the date was fixed for the end of February 1996. We began to draw up a list of those who should be invited—anyone who looked as though they were serious about genomic sequencing (and had funding) as opposed to those who were simply sequencing genes. Once they heard what was going on, others started banging on the door to be allowed in. The list included representatives from the Department of Energy sequencing labs, Mark Adams and Craig Venter from TIGR, Eric Lander from the Whitehead Institute, many others from United States labs, sequencers from the U.K., France, Germany, Italy and Japan, as well as representatives of the databases and the funding bodies, and individuals who had influenced genome policy.

The meeting, held in the Hamilton Princess Hotel under grey skies in the off season, turned out to be extremely constructive. It was the first opportunity people had had to compare notes in a comprehensive way. I think it did an awful lot of good; simply talking about the amount of sequencing that Bob's lab in St. Louis and the Sanger Centre were doing had quite a salutary effect. Those who were sequencing a few thousand bases could no longer claim to be major players, and anyone pursuing a less effective strategy was immediately exposed: the meeting had a major impact in consolidating strategies and methods. The smaller groups realized that they had either to get grants and reorganize themselves to do a significant chunk, or to accept that they were going to be gene sequencers now, and not part of the big thing. The most important thing about that first Bermuda meeting was its political aspect, sorting out who was doing what. It was there that we first began to work out what I called the 'etiquette of sharing.' We had to work together, because nobody at that stage could do the whole thing by themselves. Everyone arrived with claims on pieces of paper announcing their intention to sequence a particular region, and during the course of the meeting any competing claims were sorted out. One of the outcomes of the conference was that all these claims were recorded on a website called the Human Sequencing and Mapping Index, which was initially managed by Susan Wallace, administrator of the

United States office of HUGO. It was a logical extension of the role HUGO had established for itself of keeping track of who was doing what in the genome.

At the end of the meeting we had a session to discuss the question of data release, and this is the issue with which the Bermuda meeting is most associated in people's minds. Bob and I had always released the worm data promptly, and we continued to do the same with human sequence. But we were aware that not everyone did. There was no mechanism at the time for putting unfinished data in the public databases; they were at the time for finished sequence only. But, just as we had with the worm, we made all of our unfinished human sequence data available electronically from our own sites, so that anyone could download the information and do what they liked with it. We just asked that they recognize that it was unfinished, and acknowledge the source of the data in any publications. Everyone at the Sanger Centre, and anyone who collaborated with us, had to accept that we would do this. It was something we were doing all the time, often against opposition from those who wanted to find useful (and maybe profitable) genes before anyone else did.

I felt strongly that the principle of free data release had to be accepted, or nobody would trust anyone else. Bob and I were running the session in Bermuda, and I found myself standing there in front of a horseshoe of chairs, making my pitch. I thought it pretty unlikely that everyone would agree; several of those present, who included Craig Venter of TIGR, already had links to commercial organizations and might oppose the idea of giving everything away for nothing. But as I stood there, scribbling away on the white board, rubbing words out and rewriting, we hashed out a statement. The Wellcome Trust still has a photo of that handwritten statement, with its three bullet points. It reads:

- Automate release of sequence assemblies > 1kb (preferably daily)
- Immediate submission of finished annotated sequence

- Aim to have all sequence freely available and in the public domain
  for both research and development, in order to maximize its benefit
  to society

While Bob and I were working with our scientific colleagues,
Michael Morgan was doing the same with the representatives of the
funding agencies.

It was crucial that people from the funding agencies were able and
willing to support that policy. Bob and John needed to be able to take
the scientists along, but the scientists also needed to be driven by their
funding agencies saying, 'We ain't going to fund you if you don't
come on side.'

What I had written on the board, with minor modifications,
became known as the Bermuda Principles, which have served as a
point of reference for publicly funded large-scale sequencing ever
since. I was amazed that in the end everyone put their hands up to
this; I had no idea that it was going to go so far. Some of the delegates
agreed with the principles but had difficulties with their practical
implementation because their national governments were strongly
opposed to the requirement for rapid release, which implied that the
sequence could not be patented. 'A lot of the smaller countries did not
trust the United States,' says Michael. 'They thought they were just
pretending to be committed to free release while patenting on the
side.' But there could be no exceptions, otherwise the whole thing
would break down. Open access and early release mean that anyone
in the worldwide biological community can use those data and turn
them into biological understanding and ultimately into new inventions
that can be patented. But the sequence itself in its raw form when
publicly released becomes unpatentable. And in Bermuda, for the first
time, we won the acceptance of most (though not all) of the genome
sequencing community that this was a desirable state of affairs. It

boded well that so many people had come to share the same vision of the genome sequence as 'the heritage of humanity', the phrase adopted in the first article of the Universal Declaration on Human Rights and the Human Genome at the General Conference of UNESCO in 1997.

The Bermuda conferences, or to give them their proper name the International Strategy Meetings, became annual events. Although after 1998 they were no longer held in Bermuda, they continued to have the important dual purpose of regulating who claimed what as their sequencing territory on the genome, and maintaining the data release policy. They also began to set standards for the quality of the data that were released.

By the middle of 1996 the scene was really set for a concerted attack on the human genome. There was an international consortium, more or less in agreement and with serious levels of funding (although not as much as Bob and I would have liked). At the Sanger Centre we had teams set up, and a plan for how we would progressively increase our output of human sequence. In July we moved into our new building, gleaming and futuristic (at least compared with the old brick buildings we'd been in so far), set among trees and next to an artificial lake. Although the increased space was welcome, we immediately encountered a problem that I characterized as adiabatic expansion. What happens to a compressed gas when you give it more space? It gets cold. In the old buildings we had all been more or less on top of each other, with many of the sequencing machines in a central area nicknamed the 'goldfish bowl' surrounded by offices off a raised gallery. In the new building we all lost touch with each other as we disappeared into labs and offices separated by miles of corridors. I'm very dependent on informality, and was rather unhappy that I would not be bumping into people all the time any more as I had in our previous cramped quarters. We had to meet more regularly, to make sure we kept in touch. But I was confident that everyone knew what he or she was doing.

Later the same year the Sanger Centre shared in the celebration of another major landmark: the completion of the yeast genome, which was published in October. An international (and largely European) consortium in which Bart Barrell's group at the Sanger Centre and Bob's colleagues at Washington University had played a major role, the yeast sequencing project was the first to decipher the genome of an organism more complex than a bacterium. Apart from providing an estimate of the number of genes needed to run a cell with a nucleus, similar to those of higher animals (an article in *Science* was headed 'Life with 6000 genes') and opening the way to understanding many of the functions of such cells, the yeast sequence was a huge asset to the worm sequencing project. We could now assemble the sequence of our YAC clones efficiently because we could eliminate the sequences that belonged to the yeast. With the yeast genome all in the databases, it was now just a matter of cross-checking electronically. Nothing, I thought, could now stand in the way of getting the worm genome finished.

5 RIVALS

AS THEY GATHERED AT COLD SPRING HARBOR ON 12 MAY 1998 FOR
the annual symposium on genome mapping and sequencing, the
leaders of the Human Genome Project were in various stages of
shock, anger and despair. The previous Friday, 8 May, word had gone
out that Craig Venter had obtained private backing to form a
commercial company with the declared intention of sequencing the
entire human genome in three years. The as-yet unnamed company
would become, said Craig's publicists, 'the definitive source of
genomic and associated medical information that will be used by
scientists to develop a better understanding of the biological processes
in humans and deliver improved health care in the future.' The press
release, issued two days before the conference opened, talked of plans
to 'make sequencing data publicly available' by releasing data every
three months, but at the same time declared Craig's belief that 'this
information has significant commercial value.' I could see no other
interpretation than that Craig was aiming to gain total control of the
information contained in the genome for commercial gain. His whole
philosophy appeared to run directly counter to everything we had
fought so hard to achieve through the Bermuda agreements.

There had been almost no forewarning. Only a week before, Jim Watson heard what was up through a call from Richard Roberts. Roberts, though British-born, had spent much of his scientific career at Cold Spring Harbor working alongside Jim, and was co-discoverer of 'split genes', the phenomenon by which non-coding introns interrupt the coding sequences or exons. Since 1992 he had been based at a private company, New England Biolabs, which supplies restriction enzymes (many of which Roberts discovered) to the molecular biology community. Roberts told Watson that he was chairing the scientific advisory board of Craig's new company, gave him an outline of what was planned and asked him if he wanted to be involved. 'I wondered why Craig didn't call me himself,' says Watson, who was sufficiently worried to call Michael Morgan at the Wellcome Trust. Michael was crossing London in a taxi at the time, on the way to a planning meeting for a proposed expansion of the Genome Campus. As soon as he got the message, he rang Jim back from the taxi. 'He told me that Craig was going to make a major announcement on Monday about sequencing the genome in a year, privately—what were we going to do about it?' says Michael.

Michael had been working with us for the previous few months from the Trust's side on a renewed submission of our proposal to ramp up from a sixth to a third of the genome, to match serious United States commitments expected later that year. His first thought was that Craig's initiative would torpedo the proposal. 'I thought the governors would say, "If it's going to be done, wonderful—why should we put any more money into it?"' says Michael.

At that stage all we knew was that Craig had got a lot of money from an industrial source. Then I heard from Bob Waterston that Francis Collins and the NIH director Harold Varmus had been invited to meet Craig on Friday 8 May. It was only at that meeting, at the United Airlines Red Carpet Club at Dulles Airport in Washington, that Collins found out that the commercial partner was Mike Hunkapiller of ABI, the company that made the sequencers

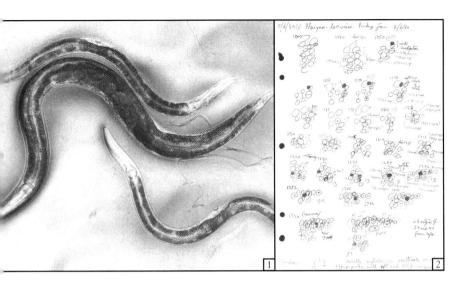

At the Laboratory of Molecular Biology in the 1970s, Sydney
Brenner's plan to trace the lineage of the nematode worm
Caenorhabditis elegans (1) became a reality. Drawings of dividing
cells (2) filled my notebooks. Bob Horvitz (3) and Judith Kimble (4)
also worked on the worm lineage and later set up their own worm
labs in the US. (5) Sydney himself (*left*) succeeded Max Perutz
(*right*) as director of the LMB in 1979.

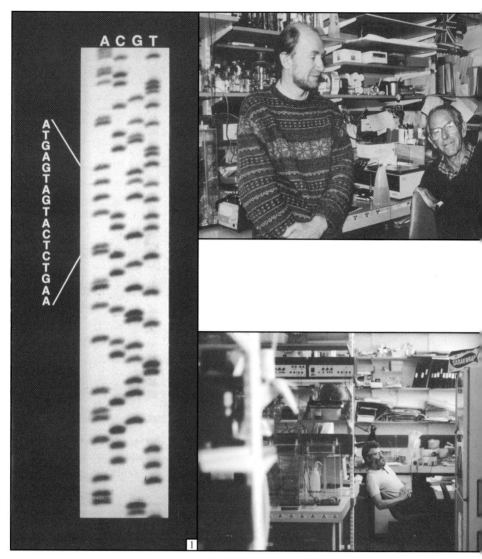

Fred Sanger invented the dideoxy method of sequencing DNA (1). (2) When Fred *(right)* retired in 1983, his assistant Alan Coulson *(left)* joined me in the cramped conditions of Room 6024 at the LMB (3), to begin the mapping of the worm genome.

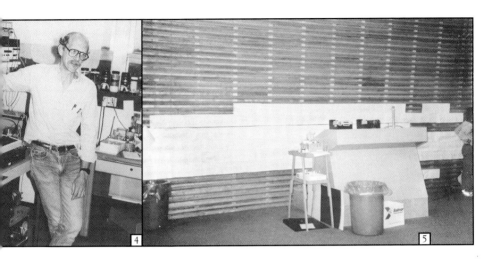

Bob Waterston (4) started to collaborate with us in 1985 and we have worked together ever since. Unimpressive at first sight, the map of the worm genome we stuck up on the wall at Cold Spring Harbor (5) prefaced the launch of international large-scale genome sequencing, celebrated by Bill Sanderson's 1990 cartoon in *New Scientist* (6). The LMB's new director Aaron Klug (7) gave us his full support.

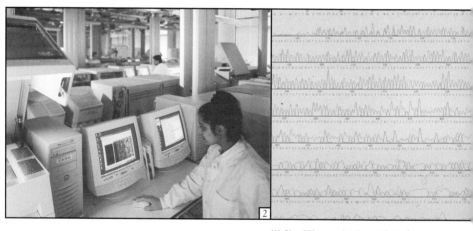

(1) Jim Watson (*right*, with Sydney Brenner) was the first head of the Human Genome Project. In 1992 the Wellcome Trust and the Medical Research Council founded the Sanger Centre at Hinxton, where rooms full of sequencing machines (2) read out human DNA twenty-four hours a day (3).

The centre was run by the board of management (4): in 2000 it consisted of (*back row*) Bart Barrell, Murray Cairns, Alan Coulson, Mike Stratton; (*front row*) Jane Rogers, John Sulston, David Bentley and Richard Durbin. At the first international strategy meeting in Bermuda, a handwritten overhead (5) outlined the principles of free release of human genomic data.

GENOMIC SEQUENCE GENERATED BY LARGE SCALE CENTRES

RELEASE

- Automatic release of sequence *assemblies* >1k's (preferably daily)
- Immediate submission of finished annotated sequence

~~~~~ and to the public domain

- Aim to have all sequence freely available for both research and development, in order to maximise its benefit to society.

POLICY

- The funding agencies are urged to foster these policies

(1) President Bill Clinton, flanked by Craig Venter (*left*) and Francis Collins (*right*), called 26 June 2000 'a day for the ages' as the draft sequence was announced. (2) In London Mike Dexter (*right*) and Michael Morgan (*centre*), fielded questions, while Tony Blair (3) shared the moment.

(4) At the hardcore analysis group meeting in Philadelphia in October 2000 we worked on the draft publication (*left to right*) Richard Gibbs, Evan Eichler, Francis Collins and Eric Lander), which finally appeared in February 2001 (5). At the 12 February Washington press conference, Eric Lander (6, *right*) explained who had done what, and how.

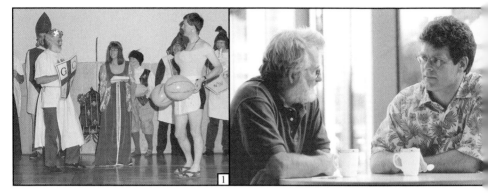

With the future of the genome secure,
I retired in 2000 and received a wonderful
send-off in the form of a Sanger Centre
pantomime (1). My successor was Allan
Bradley from Baylor College of Medicine
in Houston (2, *right*).

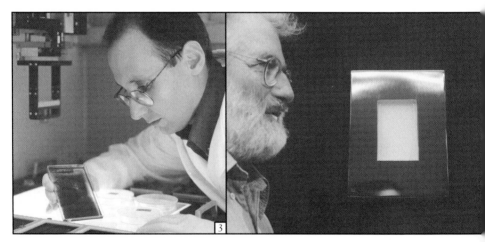

Sequencing goes on, but the centre is
shifting its focus towards biological
problems, such as Mike Stratton's work on
the genetic basis of cancer (3). Meanwhile
the genome has passed into contemporary
culture: in February 2001 the artist Marc
Quinn (4, *right*) used my DNA in his new
work for the National Portrait Gallery.

used by almost all the genome labs. Hunkapiller had been a key figure in Lee Hood's CalTech lab, where the sequencers were originally developed, and between them they had launched ABI to make the machines commercially, with Hunkapiller at its head. In 1993 the company had been bought by the scientific instrument makers Perkin-Elmer, and it was Perkin-Elmer that was putting up the money, $300 million or more, to launch Craig's new sequencing initiative. Its chief executive, Tony White, was in the process of turning a successful instrumentation company into one of the most aggressive and successful players in the high-technology sector. It had been Hunkapiller's idea to invite Craig to head a new subsidiary specializing in selling genomic information and software tools for genomic analysis.

At first I didn't fully realize the size of the challenge. I did see immediately that we were no longer going to be the largest genome center in the world. That was disturbing. Anyone who had more machines could dominate in terms of speed, not just on the human genome but on other sequencing projects. For example, after finishing yeast Bart Barrell's team at the Sanger Centre had embarked on a series of projects to sequence pathogens that caused diseases such as tuberculosis and malaria. I remember telling Bart that we would have to be on our guard. He was already used to facing challenges from Craig Venter, having successfully raced TIGR to complete the sequence of the TB bacillus. I also saw that we might have to compete on the human sequence by adjusting our own strategy to do more upfront sequencing, rather than keeping the sequencing and finishing in step as we had up to now. This was, after all, the essence of Bob's proposal of more than three years before. It wasn't a matter of racing to be first and claiming the glory. If there was a danger that private interests could gain control of the fundamental information in the genome sequence, it would be more important than ever to push that sequence out as fast as possible.

The weekend before the Cold Spring Harbor symposium passed

in a flurry of e-mails as we at the Sanger Centre and our colleagues in the United States tried to assess what our response should be. The official line from Francis Collins's office quickly emerged as cautiously welcoming. Reporting on the Friday meeting, Mark Guyer, who was Francis's deputy at the National Human Genome Research Institute (NHGRI), wrote to all the genome center heads to say, 'This is a very exciting development, providing a major infusion of resources and new technology at a critical juncture in the progress of the genome project.' In the first of what was to become a series of bizarrely staged shows of unity, Craig and Mike Hunkapiller appeared alongside the NIH director Harold Varmus, the NHGRI director Francis Collins, and the director of the Department of Energy's Office of Biological and Environmental Research, Ari Patrinos, at a press conference in Washington on the morning of Monday, May 11. Craig and Mike and some of their colleagues were to attend the Cold Spring Harbor symposium, which began later in the week, to 'encourage discussion and interaction about their plan.'

I hadn't planned to go to Cold Spring Harbor at all. I was hardly going to any meetings at that time—as far as I was concerned my most useful contribution was getting the worm genome finished, a goal that was now well within reach. Once the Sanger Centre was up and running, I had always delegated everything else: David Bentley was in charge of human genetics, Jane Rogers in charge of sequencing, Richard Durbin in charge of informatics, and so on. I took the view that they should be the ones to represent the Sanger at international meetings. Maybe that made me a rather unusual sort of director, but it emphasized the strength of the team. People were responsible for what they were doing, and I think that was recognized as a strength in the end. But after the news broke about Craig's company, Bob and others urged me to go after all. The Venter plan was a potential threat to the whole genome project, and the heads of all the publicly funded genome labs needed to formulate a response as fast as possible.

I was due to meet the Wellcome Trust governors on the Wednesday of the symposium week. As arranged months before, we would be presenting the new bid to ramp up human sequencing. Michael Morgan and I decided at once that if we got the money we should go over to Cold Spring Harbor straight afterwards and make our own announcement.

Craig did not as a rule come to the Cold Spring Harbor symposia, but ran his own TIGR-sponsored events. On this occasion there was a 'pre-meeting' for principal investigators funded by NHGRI scheduled for the day before, which as leader of the TIGR group (TIGR was sequencing part of chromosome 16) he would be expected to attend. Whatever the original agenda, the priority was now to hear at first hand what Craig was proposing to do. He duly made his case to the heads of the other sequencing labs. Not surprisingly, the reception was hostile. Not only was he going into direct competition with his former collaborators, but he had already secured a massive public relations advantage in giving the whole story to an appreciative *New York Times* journalist, Nicholas Wade, for the May 10 Sunday paper. Anyone reading that article who was not familiar with the complexities of genome sequencing would have no reason to doubt that Craig was about to pull off an extraordinary scientific feat that would leave the publicly funded project standing. At the Washington press conference the day after the article came out, he had suggested that perhaps it would be better for everyone if the HGP left the human genome to him and turned its attention to sequencing the genome of the laboratory mouse. He repeated this proposal to the stunned researchers at Cold Spring Harbor. 'It was like asking them to walk into the sea and drown,' says Jim Watson, who although no longer responsible for the project retained an understandably proprietorial interest. 'It is an understatement to say it was done in an insulting fashion.'

At his meetings with Varmus and Collins, Craig had apparently proposed that NIH collaborate with him by sharing data. If he really meant this to be a two-way deal, it might have been interesting—of course, all the HGP data was publicly available anyway. But at the same time, as the press conference and *New York Times* article made clear, he was implying that as far as the human sequence was concerned, the public project would become redundant. 'Congress might ask why it should continue to finance the human genome project through the National Institutes of Health and the Department of Energy,' Wade had written, 'if the new company is going to finish first.'

The anguish at Cold Spring Harbor was palpable. Jim Watson was so incensed that, in characteristically unguarded language, he compared Craig's attempt to take over human sequencing with Germany's invasion of Poland, and demanded to know whether Francis Collins was going to be Churchill or Chamberlain. He feared, for a few days at least, that the network of international collaboration that he had done so much to foster might be destroyed overnight. 'I worried that there would be the perception that we couldn't win,' he says.

That worry was certainly justified. On the face of it, Craig's proposal looked extremely strong. He himself was a scientist with a proven track record in running high-throughput sequencing labs. His commercial partner was the head of the company that made the sequencing machines, ABI. The joint venture was funded to a degree that was beyond the wildest dreams of any of the individual genome project labs. Large-scale genome sequencing is expensive: huge amounts of money are needed to buy the machines, hire the sequencing teams, and above all pay for a constant supply of expensive reagents—the enzymes and nucleotides used in the sequencing reactions. And the venture was founded on an important technological advance. ABI had now prototyped its new sequencing machine, the 3700, which could produce sequence several times faster than the previous model. Instead of running the DNA samples through lanes in slabs of gel, the new sequencers ran each sample through a fine gel-filled tube, or capillary, which made it possible to automate the loading of samples and reduce the amount of human intervention needed. The new company was to be endowed with 230 of these machines, each capable of reading 96 samples simultaneously. It aimed to have an output of 100 megabases of raw sequence per day.

There was absolutely no doubt that, armed with these resources, Craig was capable of equalling or even exceeding the existing world output of raw genome sequence, assuming that the as yet untested machines performed as they should. But as Bob and Eric and the other genome leaders questioned him about the details of his strategy, they became convinced that it just wouldn't work. Craig was proposing to speed the process up by using the whole-genome shotgun method he had employed to sequence bacteria such as *Haemophilus influenzae* and *Helicobacter pylori*. Instead of mapping clones by fingerprinting and shotgun sequencing them one by one, as we had done for the worm and were doing for the human, he was going to shotgun the whole genome at once.

It wasn't the first time someone had proposed to do this. At the Bermuda meeting in February 1996 James Weber, a human geneticist from the Marshfield Medical Research Foundation in Wisconsin, argued that the search for human disease genes could be greatly speeded up by the whole-genome shotgun approach. In partnership with Gene Myers, a computer specialist from the University of Arizona in Tucson, he followed up the presentation with an article in the journal *Genome Research*. But the genome project scientists rejected the idea. In a closely argued response to their article, Phil Green of the University of Washington, who knew as much as anyone about the problems of assembling human sequence, pointed out a number of shortcomings with the proposal. The most serious, from a scientific point of view, was that the human genome was a very different proposition from the bacterial genomes on which the strategy had previously succeeded.

155

Bacterial genomes consist of about 2 million bases and the human genome 3 billion. But they're also qualitatively different, especially in terms of the amount of repetition there is in the genome. Bacterial DNA typically includes about 2 percent repeat sequences; human DNA includes about 50 percent. Repeat sequences mean that one DNA fragment might have matching overlaps with many other fragments. Only by working with fragments taken from larger clones whose position in a whole-genome map is already known is it possible to be sure you've assembled the whole sequence correctly. Furthermore, some regions of DNA are much harder to clone in bacteria than others, so a random approach leaves many more gaps than you might predict on a purely theoretical basis. One of the reasons Bart Barrell's group was able to complete the TB sequence ahead of TIGR was that the whole-genome shotgun approach had not worked in that case. TB is exceptional among bacteria in having a lot of repeats that are also hard to clone.

Craig might have managed to gather enough computer power and software expertise to solve the assembly problem through sheer brute force. But we wouldn't know until it was too late. Our method, mapping first and then sequencing the mapped clones, though laborious at first sight, meant that you could check how good your data were as you went along. The quality of the product was guaranteed, and our priority had always been to produce a sequence that would stand for all time. There was no doubt that the whole-genome shotgun would deliver plenty of raw sequence, but how many pieces would the assembly end up in? If it turned out to be full of holes, it would have some use for gene hunting; but it would be prohibitively expensive to finish, and the location of the smaller pieces on the genome would be unknown. Phil Green had made all these points in print the year before, and yet Craig was going ahead anyway.

But the greatest concern was over exactly what he intended to do about releasing his data. Officially they were to be released 'on a quarterly basis.' But it was difficult to see how the company could succeed as a commercial supplier of genomic information if it did indeed make all its data public in this way. The initial announcement said that it would be seeking patent protection on 'only 200–300 genes'—not a very competitive position when other companies, such as Incyte and Human Genome Sciences, were already claiming rights to many more. So while the HGP leaders were doubtful about Craig's science, they were even more doubtful about his business plan. It would be all too easy for him to backtrack on his early commitments in order to satisfy his shareholders. 'It was a testy meeting,' says Bob Waterston. 'We asked Craig hard questions, and he thought we were just being nasty to him. And we might have been a little nasty, but scientists are skeptics. Craig has always been very thin-skinned that way—he reacts very badly to criticism.'

Craig had one answer to critics of his scientific strategy: he announced that he was going to test the method on a smaller organism first, and then, within the hearing of many of the conference participants, asked Gerry Rubin to step outside and have a word. The team Gerry had assembled in Berkeley, California had begun to sequence the fly genome in 1995. By 1998 they had done 20 megabases out of 120, and were planning to scale up to finish the rest, with the help of Richard Gibbs who now had a large sequencing center at Baylor College of Medicine in Houston. As they saw the burst of discoveries that followed our steadily increasing output of the worm sequence, the much larger fly research community was clamoring for the fly genome to be completed faster.

Gerry just wanted to get the fly done—he had already put more time into it than he wanted to, having taken it on in the first place largely because nobody else looked like doing it. So when Craig told him that his company was going to sequence the fly and would like to do it as a collaboration, he was prepared to listen.

He didn't know whether I was going to belt him or whatever, but I immediately said 'Great, anyone who wants to help finish *Drosophila* is my friend, as long as you are going to put all the data in GenBank.' And he said 'Fine, I'll do that.'

Gerry is a pragmatist; he saw immediately that he could not compete with Craig and that it would be to his advantage to throw in his lot with him. I talked to him about what had happened and agreed that he had no choice. A few months later Celera—as Craig's company was eventually named—formally signed a memorandum of understanding with the University of California (the official recipient of the NIH grant which funded Gerry's sequencing work), including the commitment to release the data freely.

It was surprising to a number of people that Gerry agreed so readily. Scientists in the academic world stake claims to little plots in the field of scientific inquiry which others by and large respect. Although there are plenty of examples of 'races' to make new discoveries and arguments over priority in the history of science, the processes of grant distribution and publication make them much less frequent than one might expect. True, if someone doesn't seem to be getting anywhere then others might move in and take over. But no publicly funded scientist, in the United States at least, would have applied for a grant to sequence the fly genome because everyone knew that Gerry was doing it. The NIH would not have funded such a grant anyway because funding bodies don't like duplication. Part of the reason why many people find Craig hard to stomach, and why others admire him greatly, is his cavalier disregard for such academic niceties—and with Perkin-Elmer's money behind him, he could afford it. But Gerry says he had never felt possessive about the fly genome project; his little plot was in fly molecular genetics, and he wanted to get back to learning how flies are put together as soon as possible. He kept a small genomics program going in his lab in case Craig's effort failed, but otherwise agreed to hand responsibility for the shotgun part of the sequencing effort over to the company.

Craig was too late to take control of the worm. Back in 1990 labs had been taking bets about which would win, fly or worm, and at that time we were not the favorites. But once Bob Waterston and I got going, the worm opened up a huge lead; by the time of Craig's announcement it was almost done. If we had not been so far ahead, he might just as easily have picked on the worm; or if he had not, the fly would certainly have caught the worm up once he took it on. It was good that, at Bob's insistence, we had gone faster with the worm sequence than was originally planned. If we had gone at my pace and not Bob's, Craig would have been able to catch us up. Why did it matter to be first? It's not just a question of personal pride. Our labs by this time were large-scale enterprises, employing a lot of people and with a great deal of hard-won money invested in them. We had to set ourselves tough goals and meet them in the face of competition in order to justify the investment that had been made in us, and keep the credibility that would ensure our future funding. These are the forces that push science forward efficiently. In addition, our scientific credibility gave us influence over the quality of the product and how data were handled.

Craig had declared his intention of invading the fly genome and looked like taking the flag without a fight. Would he be able to do the same with the human? He had certainly opened up a huge public relations advantage, something that was to happen time and again in the years to come. I'm sure he has good advisers, but there's no question that his own instinct for the timing and presentation of announcements, and for the apparently casual aside, is superb. Again, it's a skill that most scientists never learn and rarely have to practice. The convention is that you don't announce anything until the work is completed and accepted for publication in a peer-reviewed journal. That way there should be no embarrassing retractions after hasty announcements, as happened in the case of Stanley Pons and Martin Fleischmann's 'discovery' of cold fusion in

1989. But Craig was no longer in science, he was in business. And the priority for a business is not scientific credibility but share price and market penetration.

The initial press reports on Craig's company, beginning with Nicholas Wade's *New York Times* article, set a trend for presenting the new venture as a more effective alternative to the Human Genome Project. Just to take two examples, it was said that the new company would sequence the genome faster, and would do it more cheaply. The initial Perkin-Elmer press release said that the company would have a 'substantially complete' genome in 2001, four years ahead of the Human Genome Project's target of 2005. It did not mention cost, but most articles quoted the figure of around one tenth of the $3 billion budget of the HGP suggested by Craig at the 11 May press conference. But these reports weren't comparing like with like. The public project's target date was for finished sequence, with all possible gaps closed and to an accuracy of 99.99 percent. Craig's company, as he made clear to the genome center leaders at Cold Spring Harbor, did not plan to bother with the time-consuming process of closing gaps, which takes a lot of manual intervention, but aimed to get as much as he could out of a fully automated procedure and then stop. Although Craig later insisted that his sequence would be 'highly accurate' and 'comparable to the standard now used in the genome sequencing community of fewer than one error in 10,000 base pairs', it was clear (as subsequently turned out to be the case) that the whole-genome shotgun method would produce what we now call a draft, not a fully finished sequence.

As for the cost, the Human Genome Project funded far more than human sequencing. Apart from other genomes, such as worm, fly and yeast, it had a program on the ethical, legal and social impact of genome research and a number of projects on the development of new technologies, and supported databases and bioinformatics. Craig also said at the press conference that he could get the cost of sequencing down from the genome centers' then figure of 50 cents to 10 cents a base. But that cost, which the best centers were already undercutting, was for finished sequence, and it's the finishing stage that accounts for at least half the cost: on this basis, the target of 10 cents was the one Bob Waterston and I had suggested two years earlier. Another widely quoted criticism of the HGP was that halfway into its fifteen-year schedule it had sequenced only 3 percent of the human genome; but we had deliberately spent the first six years scaling up through smaller genomes and developing the techniques. As the HGP's founders had planned in the 1980s, the whole field had moved forward to the point where tackling the human genome was feasible, when earlier it had not been. Indeed, our purchases from ABI, and collaborations with them, had helped to drive the development of the faster sequencing machines that Craig's company was planning to use—and which we too had every intention of using. Finally, that figure of 3 percent represented only finished sequence. Taking into account the assembled but unfinished clones that were in the databases, the figure was 13 percent.

As head of the National Human Genome Research Institute, Francis Collins had to take the brunt of the overt and implied criticism of the genome project. At the Sanger Centre we had the luxury of knowing that we had a politically independent body, the Wellcome Trust, behind us all the way. But in the United States, with a Republican Congress, Francis feared that any suggestion that the private sector could do a better job than government-funded labs was going to find a receptive audience.

> The public project was portrayed as laboring with a clumsy, bureaucratic, difficult-to-implement strategy, and these fast-moving folks in the private sector were going to run circles around us with their fancy whole-genome shotgun approach. And that was really quite unjustified and hurtful.

It was also potentially a serious threat to the very survival of the publicly funded project in the United States.

Trying to get reporters to print the admittedly more complex analyzes that we felt were being ignored was going to be an uphill struggle. We were learning fast that we would have to play the public relations game if we were to survive. But that didn't mean indulging in empty hype. What we needed was a big vote of confidence in the public project as a counter to Craig's hints that it was an expensive white elephant. And that's exactly what Michael Morgan and I hoped we would get from the Wellcome Trust governors that Wednesday. Long before we had any hint of Craig's company, we had carefully prepared our case for doubling our output. Jane, David and Richard had all put together detailed accounts of how it would be done, with breakdowns of the cost—all the stuff that I can't handle. I went up to London with Jane and Richard in the train. In my bag I had one handwritten overhead, containing a few bullet points; I scrawled an extra one about Craig's company at the bottom.

Seated round the table in the panelled boardroom at the Trust's offices in Euston Road on May 13, the scientific governors of the Wellcome Trust waited expectantly. I spoke for only a few minutes. I told them that I thought we had a very strong case, and that the U.K. could not continue to play a leading role in the Human Genome Project unless we took on more than one-sixth of the work. Then I added that the arrival of Craig's company made it crucial that we should maintain a strong presence, if genomic information were to be made freely available to all. To the question 'Why should *we* do it?' I had a ready answer. Three years before it had looked as if other countries outside the United States, particularly France, Germany and Japan, were prepared to take a serious stake in the project. But by 1998 it was clear that this was not going to happen; they would only ever be minor players. If there was going to be an influential international stake at all, it would have to be taken by the

U.K. And without a strong international presence, there was a real danger that the whole project would fall into the hands of the commercial sector.

We went outside to wait for their decision, and soon afterwards the message came back that they had agreed: our funding would be doubled, giving us the resources to take on and finish a third of the human genome. Once the governors understood that Craig Venter's initiative was essentially a privatization of the genome, Michael Morgan felt there was no risk that they would pull out on us.

By the time we got to that Wednesday meeting, everyone's dander was up. The reports were all very positive, and the governors just said, 'We must do this.'

It was an emotional moment. I sent a message back in to the governors thanking them for their support, and saying that I saw us all as joining together in a partnership to meet our sequencing goals. What I didn't say was that with the new grant we felt the governors had shown their trust in us to an extent that we had often wished for but not always experienced in the past. Yes, they had invested hugely in us, but somehow there had always been a feeling that they didn't quite believe in us. The loss of the MRC's support had made things very difficult, and maybe no more could have been done anyway, but I still felt the Trust had missed an opportunity in not giving us the green light to go for a third of the genome in 1995. For the next two years there was a feeling among us that the Sanger Centre was just not doing what it was designed to do. We did a lot of things, and we were successful, but we were forced to mark time to some extent. We offset this as much as possible, though, by scaling up where we could and committing resources in readiness. Richard Durbin says now that I drove the scale-up very strongly, claiming extra chromosomes, moving worm sequencers across to human as the worm neared completion, and 'engaged in justifiable brinkmanship.' I don't recall doing much driving, but I certainly made no attempt to rein it in.

With the new grant, all that frustration could be put behind us. Michael was full of excitement—the first thing he did was open a bottle of champagne, and then he and I immediately sat down in the press office with Jane and Richard and drafted a press statement. Jane flew to New York that evening, and Michael and I followed the next day.

It was difficult to sustain our euphoria that first evening at Cold Spring Harbor. Everyone else was very down. Craig had left before the symposium proper even began, tight-lipped at the criticism he had received but in no mood to compromise. What was needed was a robust, coherent response, but there was little agreement about what that response should be. Broadly speaking, people had divided into two camps. One side (most emphatically represented by Eric, who on this occasion was right) argued that we should immediately change our strategy and speed up the production of sequence to beat Craig at his own game, and worry about finishing later. The other (including a number of those in the Sanger Centre and Bob's lab in St. Louis) favored the view that we should not be bounced into such a major change of direction, but stick to our existing strategy of sequencing and finishing in step. There was pretty much agreement on two points: that however we got there, our ultimate goal was still to produce 99.99 percent accurate finished sequence, and that, as far as the American side was concerned, we stood in real danger of losing everything if, as we feared from the tone of his press statements, Craig were to persuade Congress that it was wasting its money.

Weary and jet-lagged, I sat with Bob and his group from St. Louis in the Blackford restaurant, anxiously going over our dilemma. The group was strongly against a change in strategy, but feared that going at the same pace would allow Craig to make political capital out of the HGP's 'slow progress.' Rick Wilson said gloomily, 'This is a lose–lose situation.' Earlier, almost as soon as I arrived, Jane had grabbed me, saying, 'You've got to do something!' There was a real possibility that the public sequencing effort in the United States would collapse, but even if it wasn't that bad she feared that our tried and tested map-based methods might be under threat if those of our own HGP partners such as Eric Lander, who had less to lose from a shift to a less systematic approach, won the day. Eric seemed more upset than anyone about Craig's announcement, and at one point he had a furious argument with David about mapping. I found all this very puzzling, because at this point he didn't appear to have much of a stake in the outcome. It was not until much later that I finally understood that he was positioning the Genome Center at the Whitehead Institute to be the leading producer of raw sequence data.

Throughout that first evening I caught grins on the faces of symposium participants who were not involved in large-scale sequencing as they saw the members of the genome labs huddled grimly in corners. It was quite clear that they were enjoying the thought that this very small group, who had taken so much in funding to produce the human sequence, were finally getting their come-uppance. And even those who weren't laughing did not really understand what we were up against. Mike Smith, who headed a genome center in Vancouver, wandered into Bob's meeting obviously expecting to join in with a bit of academic debate as one normally would in the breaks at the symposium. I told him it was a private meeting, and he was clearly upset. I sat with him at breakfast the next day and talked it over. The point is that the sequencers were no longer running traditionally structured labs, with a group of more or less independent scientists and a few technicians in support; we were effectively running 'businesses.' Bob and I had the biggest businesses at the time; Eric Lander aspired to have the biggest business. It was not just a matter of being on your own and so able to chuck everything out of the freezer and start again if a line of research didn't work out—which is the way you should do science. We had got ourselves in a position where we had highly trained staff now numbering hundreds. You can't treat an organization like that as a personal research group. It looked as if we were just being pig-headed and defending our interests, but we were in a position of responsibility in more ways than one: without us, the human genome would be privatized.

I tried to explain this to Mike Smith, but I don't think he understood. 'You're not leaving anything for anybody else,' he protested. But if the project had been left to all the small groups, Craig would have just walked all over them. We were facing huge pressures, and the sense of what was right and what was wrong was quite different from the normal senses of right and wrong that you might come across at biological meetings. Mike, who tragically died of cancer in October 2000, was a wonderful man, but Bob and I had been forced to adapt from his way of thinking in order to make things happen. And at that time the most important thing was to maintain morale in the lab, talk through what the Venter announcement meant for us and think about strategy.

My feeling at the time was that we didn't need a radical change of strategy, but that we should scale up and increase efficiency. And unlike the Americans, we at the Sanger Centre knew that we had the money to do it. The timing could not have been better if the script had been written in Hollywood. On the Friday morning, the symposium organizers gave Michael Morgan and me a slot before the scheduled program to make our announcement. Grace Auditorium, the lecture theatre on the Cold Spring Harbor campus, was packed. Michael got up to speak first. He explained that, in funding the Sanger Centre to double its output, the Wellcome Trust was reaffirming its commitment to an international initiative to produce a high-quality, finished product that would be publicly available for all to use. Alluding to the Venter–Hunkapiller proposal to patent some sequences, he threw out a challenge: the Trust was opposed to the patenting of basic genomic information, and would be prepared to contest such patent applications in the courts. It was just what everyone needed to hear. When he had finished, the crowd in the packed hall rose to its feet in acclamation. Michael has a great sense of theatre and knew exactly what he was doing: putting pressure on the National Institutes of Health to increase its backing for genome sequencing. He really enjoyed the moment. 'The atmosphere was electric,' he says. 'It was fantastic, very exciting.'

I stepped up straight afterwards to add a few words. I suddenly caught sight of Jim Watson at the side of the auditorium, his eyes glinting in the half-light. It immediately brought back to mind a scene in the BBC film *Life Story*, about the discovery of DNA. In a beautifully lit shot Jeff Goldblum as Watson is at the back of an auditorium watching a lecture on DNA. Knowing Jim was there, I ended with the slightly trite but nevertheless appropriate comment that whatever happened, it was now more rather than less likely that the sequence would be completed in the lifetime of a certain individual. And the crowd erupted again, stamping their feet in approval.

It was a very significant development indeed. Francis Collins called Wellcome's decision 'a shot in the arm', but Jim Watson goes even further: 'It was absolutely critical, psychologically,' he says. In Bob Waterston's view it made it possible for NIH, which had initially been too timid in its response, to 'stand up a little straighter.' After that I heard less defeatist talk about Congress pulling the plug on the HGP. The next editorial in *Nature* reported that 'The talk was of healthy competition rather than throwing in the towel.' The question was, how should we proceed? Craig's sudden appearance on the scene forced us to look for a strategy that would get visible results quickly without compromising ultimate standards.

The first opportunity for the HGP leaders to put their heads together came when we gathered at Airlie House in Virginia, an NIH conference center, at the end of May. The purpose of the meeting was to sketch out the next five-year plan for the project, covering the years 1998–2003. Richard Durbin and I both went to represent the Sanger Centre, and Michael Morgan was there too for the Wellcome Trust. Mike Hunkapiller was there representing ABI and, by implication, Craig's company. It was a tense meeting. Harold Varmus made it clear that there could be a lot of extra money available for the United States centers, but that grants would depend on an agreed strategy. It was obvious that the strategy would involve some form of acceleration, and some labs were in a much better position than others to rise to this challenge. And there were still important differences of opinion about whether we should accept the pressure to go for a fast intermediate product—a draft—or whether we should continue on our steady course to the finished sequence.

By the end of the meeting there was more or less an agreement, though with strong dissenting voices (mainly those of Richard Durbin and Phil Green), to try to get rapid shotgun coverage of mapped clones covering the whole genome—perhaps threefold coverage rather than the sixfold that was usual for the shotgun stage—at the same time as accelerating our production of finished sequence. Another objective I had was to make people realize that we had to take the threat of Craig Venter's PR skills seriously, and actively promote the public project. Towards the end of the meeting, one balmy evening by the pool, Bob and I cornered Francis Collins, Michael Morgan and Alan Williamson and won their verbal agreement (as I thought) to run a publicity campaign emphasizing the positive points of the public project and correcting disinformation from the other side.

Bob and I were both fairly comfortable with the decision to go for a fast shotgun, or draft. Since the days of the worm, we had always been accustomed to the idea of releasing unfinished data. But we both met opposition within our own labs—in my case, Jane and particularly Richard were unhappy about the idea of decoupling sequencing from finishing. And there were still major unanswered questions about how the whole job would be shared out among the partners in the international consortium.

Before the project's leaders reached a final decision the United States Congress, or at least one of its subcommittees, called a hearing to investigate how the launch of Craig's still unnamed company might affect the federally funded Human Genome Project. So, on 17 June, Craig lined up to address the Subcommittee on Energy and Environment of the House of Representatives Committee on Science, along with Ari Patrinos of the Department of Energy, Francis Collins from the National Human Genome Research Institute and Maynard Olson from the University of Washington.

Maynard has in many ways acted as the conscience of the genome project in terms of ensuring the quality of the final product. He was one of those who publicly doubted the wisdom of the 'megalomaniac genome project' when Bob and I first suggested it (see chapter 4), on the grounds that it might reduce the chances of a fully finished sequence ever being produced—although ultimately he supported us. Not one to rush into unconsidered judgements, his view of Craig's announcement, read out to the assembled members of Congress, was damning.

> The excitement generated by the well-orchestrated public relations campaign surrounding the Perkin-Elmer announcement should not disguise that what we have at the moment is neither new technology nor even new scientific activity: what we have is a press release.

He went on to issue the standard academic challenge to a doubtful claim: 'Show me the data.' And he predicted that Craig's whole-genome shotgun method would encounter 'catastrophic problems' at the assembly stage, with 100,000 'serious gaps' in the sequence. (Three years later he turned out to be spot-on.)

Francis echoed this criticism in his own presentation, and made an unfavorable comparison between Craig's stated policy of releasing his data quarterly and the daily release policy of the public project. 'Any delay can result in wasted effort in research,' he said. He also pointed out, presciently as it turned out, that 'the private effort's commitment to data release might diminish over time, if business pressures came to the forefront.' But he made it clear that 'the private and public genome sequencing efforts should not be seen as engaged in a race,' and that both his institute and the Department of Energy would welcome the opportunity to collaborate with Craig's company.

When it came to Craig's turn, he gave the committee a quick resumé of his past scientific record, including the claim that he and his colleagues at the NIH had 'developed a new strategy for identifying genes' (something of an overstatement as we saw on p. 108) before going out of his way to praise the achievements of the HGP. Moving on to talk about his company, he first assured them that 'an essential feature of the new company's business plan is to provide public availability of the sequence data.' Addressing concerns about patents, he went on to make the claim that 'Our actions will make the human genome unpatentable,' but added that his company would seek intellectual property protection on 'fully characterized important structures' amounting to 100–300 targets. Arriving finally at the issue the committee was meeting to address, he said that the impact of his company on the HGP should be 'to reorient it sooner to move beyond DNA sequencing into the research that will help us better understand and treat . . . diseases.' He also dismissed the idea of a 'race', saying that the public project should be judged 'by its ability to adapt and work with new initiatives rather than compete against them.'

I found it impossible to accept Craig's words at face value, but somehow he emerged with his credibility intact. Maynard Olson's clear-eyed criticisms, all of which turned out to be accurate, were interpreted by many commentators as sour grapes. On the other hand, nothing the congressional committee heard that day appeared to persuade its members that it was time to stop supporting the work of the NIH and DOE on human genome sequencing. Francis had been able to show that, far from dragging its feet, the public effort had met all its targets to date on time and within budget, and that it expected to meet its new targets just as efficiently.

After three months in which we had to refer to Craig's company as 'the Venter–Hunkapiller proposal' (usually shortened to Venterpiller or VentiPEde), it finally found a name, Celera Genomics Corporation, and a catchy slogan: 'Speed matters. Discovery can't wait.' The company set itself up in Rockville, Maryland, next door to TIGR's laboratories, in two block-like buildings, one for the computer center and one to fill with the new 3700 sequencers once they started to arrive. It was time for Francis Collins to seize the moment and make a statement as bold as those already made by Craig and by the Wellcome Trust. After much negotiating, he finally came out with a statement of research goals for the next five years in mid-September. The main points were that the international consortium would produce a 'working draft' sequence covering more than 90 percent of the total at an accuracy of more than 99 percent by 2001; and that we would go on to complete the fully finished sequence, with fewer than one error in 10,000 bases and no gaps, in 2003, fifty years after the discovery of the DNA structure. 'No-one else is doing this,' said Francis pointedly.

I was happy with the statement as far as it went, but under its smooth surface cracks were appearing that threatened to blow apart the fragile consensus that Francis had brokered among the HGP participants. What Craig might or might not do or say was no longer the immediate problem. The issue was partly technical and partly political. Our sequencing method depends on having a supply of mapped clones—fragments of DNA from known positions on the chromosomes that are cultivated in bacteria. Pieter de Jong at the

Roswell Park Cancer Institute in Buffalo, New York State, had produced libraries of fragments of about 150,000 bases in BAC clones
from the whole human genome, which it supplied to sequencing
centers. At the same time an ingenious technique called radiation
hybrid mapping, developed by David Cox at Stanford in the United
States and Peter Goodfellow in the U.K., made it possible to order
markers along the genome very rapidly. These markers could then
be used as probes to pick BAC clones for sequencing. David
Bentley's team of mappers used chromosome-specific probes to pick
clones belonging to the chromosomes we were working on, and
would then 'fingerprint' each clone, just as we had done for the
worm map. As long as you've started with enough clones you can
work out their order along the chromosome by comparing overlaps
in the fingerprint patterns.

Now you're in a position to start sequencing. The most efficient
way to proceed is to choose a set of BACs that overlap with each
other all along the chromosome, but only just—we call this a
minimum tiling path. Those BACs are broken down again, this time
into pieces about 2,000 bases long, ready to go into the sequencing
reactions. At the Sanger Centre we had declared our intention
of sequencing chromosomes 1, 6, 9, 10, 13, and 20, and parts of
22 and X, making a third of the genome in total. (As part of the
'brinkmanship' that Richard Durbin accused me of, we had made
some of these claims even before we knew we had the money to
scale up our sequencing—it had the effect of getting others to focus
on the necessity of ensuring that all the chromosomes would be
efficiently taken care of.) At the various Bermuda meetings others
had also staked claims to chromosomes or parts of chromosomes
by mutual agreement. Not all of these claims were backed up by
funding or capability, but there was an agreed set of guidelines to
determine what should be done if someone claimed more than they
could deliver. By September 1998, pretty much everything was
spoken for. But not everyone who had a claim to a chromosome

was in a position both to map it quickly and to gear up for high-throughput sequencing. And anyone who wanted to get started on sequencing was going to have to wait for someone to produce mapped clones.

Eric Lander was very unhappy about this situation. He was convinced that Craig was indeed capable of delivering the human sequence, and believed that nothing less than a commitment to sequence at the same speed or faster would be an adequate response. The Whitehead's sequencing center had begun large-scale sequencing work only in 1996, as part of the NIH pilot program, with a project on chromosome 17. More seriously for Eric, it did not have a program for producing a supply of mapped clones that would provide enough raw material to sequence at the rate he now wanted to go. Eric decided that the only solution was to abandon the regional, chromosome-by-chromosome approach to sequencing. Instead, he argued that anyone who had the capacity should be allowed to sequence BAC clones chosen at random from a library of clones covering the whole human genome. He explained his thinking to Francis Collins and Bob Waterston at an informal meeting in Washington in early October. As the first 100 reads from each clone came off the machine, he suggested, you would check them against a central computer database to see if another lab had already covered that clone. If so, you'd chuck it out and start another one. The sequenced clones would be arranged in their correct places on the chromosomes by mapping them afterwards.

Bob was very disturbed by this conversation. It was clear that it was not the first time that Eric and Francis had talked about the plan, and it looked as though other United States labs were also interested. The most disturbing aspect of it was the implied shift in strategy away from the regionally based approach—as in regions of the genome—that had been worked out at the Bermuda meetings, and which seemed to have been accepted at Airlie House just four months earlier. Sequencing clearly mapped regions provides the best guarantee of the quality of the final product, which is what we were supposedly all trying to achieve. But the approach also made it possible for the project to work as a genuinely collaborative international consortium, with the work shared out among the participants according to their interests and their research capacity. It was open and honest, with everyone knowing what everyone else was doing. Sequencing random BACs, in contrast, left the door open for those who might want to 'cherry-pick' BACs that contained important genes, so that they could get a head start in pursuing commercial developments. And while our regionally based commitments included every step right through to finishing, it was not clear, if everyone scrambled to put out as much sequence as possible from random BACs, who would be responsible for finishing. Eric was apparently unconcerned about this, telling Bob that we could 'worry about the details later.'

My immediate reaction when I heard about this plan from Bob was that it represented a direct threat to the non-United States members, and also to Bob himself who worked in much the same way that we did. If Eric received the funding for a massive increase in his sequencing output, and if he and other United States labs were able to sequence any BAC that hadn't already been sequenced elsewhere, they would disrupt our well-organized tiling paths and destroy the efficiency of the process. Our only recourse would be to join in the random free-for-all, which would certainly produce a lot of sequence quickly, but would also present a whole new set of problems for filling the gaps afterwards. And as for the much smaller German, French and Japanese groups, who had enough problems convincing their governments that they should be funded at all, I could not see how they would survive.

Michael had not been copied into the correspondence on this issue over the preceding few weeks, and was furious that such a major change in strategy was being discussed without consulting the Wellcome Trust. NIH and Wellcome had previously agreed that each funding body would manage the work of its own centers. He wrote to Francis to complain that the agreed strategy was about to be replaced with an 'uncoordinated, expensive and self defeating free for all.' Jane Rogers, David Bentley and Richard Durbin were equally alarmed; as the individuals who dealt from day to day with our sequence production and information management, they knew how seriously a major change of strategy could hold up our operation. We made it clear that we saw no reason to change our strategy, but found that our objections were brushed aside as Eric's proposal gained support. The United States side was by no means unanimous—Bob Waterston was still unhappy about the proposed shift, and Richard Gibbs at the Baylor College of Medicine came down in favor of continuing the regional approach—but it was quite clear that Eric, now backed by Phil Green (with some reservations about the detail) and Maynard Olson in Seattle, had convinced Francis of the need to change course. Their evidence was a theoretical calculation showing that although you might end up sequencing more clones than were necessary with the mapping approach, the costs would be no greater. Towards the end of October most of the United States leaders, some more enthusiastically than others, were beginning to talk about a 'hybrid strategy', with some groups working regionally and others randomly. It sounded like a mess, and it wasn't a compromise I could support.

It seemed to me that Francis was no longer listening to other views. I felt very strongly that to relinquish 'our' chromosomes would greatly diminish our influence on the project as a whole. Keeping our chromosomes was about sticking to a scientific strategy that we knew was working in practice, as opposed to one that had so far been developed entirely on paper. We were repeatedly told in the conference calls that both strategies were equally risky. This was simply not true, since we had already sequenced the worm and large tracts of human by mapping first. Crucially, for efficiency the random approach would require the BACs themselves to be distributed evenly, and this was unlikely to be the case. But it was also about sharing control. That sounds like self-interest, and of course it was, as far as the Sanger Centre was concerned—but, more importantly, I saw the need to protect the international dimension. With a strong international partner the NIH/DOE project could stand up to political pressures to compromise with the commercial sector. Without that partner, it was not at all clear that the whole human genome might not fall into commercial hands. I certainly believed this to be a real danger, and I thought the costs to humankind would be incalculable.

We had a chance to discuss all this at the annual retreat of the Sanger Centre board of management in the middle of October 1998. It was held for the second time at Blakeney, a beautiful place on the north Norfolk coast: a tiny harbor at the edge of a broad estuary that dries to sand and mud at low tide. Huge flocks of birds live there or pass through as migrants, and you can walk out along the low sea wall to get close to them. It's a wonderful, peaceful spot, where you can look beyond day-to-day detail and see problems in perspective. The directors of the Sanger Centre's management company Genome Research Ltd joined us: Martin Bobrow, who had moved from Guy's Hospital to become professor of medical genetics at Cambridge University, and Alan Munro, the Master of Christ's College, Cambridge. Michael Morgan also came for part of the time.

Rather to my surprise, there was general agreement that the problem might be real: that I was guilty perhaps of mild paranoia, but not without cause. We were conscious of the fact that we were up against powerful economic forces, and that Francis and his advisers had dual loyalties. Under the 1980 Bayh–Dole Act, the National Institutes of Health had a right and a duty to support United States industry by encouraging the commercial development of federally funded research. At Bermuda, meanwhile, they had agreed to collaborate internationally and altruistically. There was no doubt of their desire to collaborate, but there were clearly great pressures on them; they would surely be looking for a workable compromise, in which the random approach might just be the first step. We in the U.K. had stuck our necks out for internationalism: we were contributing nothing directly to the U.K. economy by our activity (though in the long run our high-profile activity brings indirect benefits). In France, Jean Weissenbach had won a legal battle for French government funds to deliver one chromosome for the HGP. The German government, by contrast, had concluded that its best course was to compete on patentable applications, and phase out its contribution to the Human Genome Project. And indeed our own government had been quite limited in its support. So it was incumbent on us either to ensure that the project remained truly international, or else to admit that the skeptics were right and withdraw from the partnership altogether. As Richard put it, we needed to smoke out what was really going on.

When it became clear that neither Bob Waterston nor Richard Gibbs nor any of us at the Sanger Centre could dissuade Francis from adopting one or other variant of the random approach, I decided that the only answer was to put aside polite scientific discourse and let him know what I really felt. Under the heading 'Friendly fire', I e-mailed Francis to express my 'deep concern' that he was in danger of accomplishing what Craig Venter and Mike Hunkapiller had failed to achieve: the destruction of the international public program for sequencing the human genome.

I went on to accuse him of orchestrating a U-turn in strategy, and made it clear that I thought his motives were political rather than scientific (meaning that he wanted to strengthen the United States contribution to the project at the expense of his international partners). I asked whether the National Institutes of Health was really interested in pursuing a strong, international program, pointing out that on at least one occasion its press releases had somehow forgotten to mention that the genome project was not an entirely United States operation. I told him that if we could not work round the new strat-

egy we might have to consider going it alone, and implied that as far
as the Wellcome Trust was concerned, it should certainly be wary of
collaborating with the United States funding agency in future. I concluded with the sentence: 'I now know something of the feelings of
the British tank commander in the Gulf when he watched his crew
being destroyed by the weapons of his allies.'

Looked at in the cold light of day, the missive reads as an
emotional outburst, which certainly is an accurate reflection of my
feelings at the time. But it was not fired off on the spur of the
moment. It was drafted over several days, with input from close
colleagues at the Sanger Centre (most of whom made me tone it
down). I deliberately sent copies to the other genome center leaders
and to other interested parties including Jim Watson.

The fallout was immediate. Francis was personally very dis-
tressed. He was completely taken aback by the strength of my
feelings, and to be honest I had not taken much part in the e-mail
debate that had gone on over the previous month, although Jane,
Richard and David had made our position perfectly clear on the
scientific issues. But he immediately began to look for ways to build
bridges, which was exactly what I had hoped for. I was very willing
to apologize for the hurt I had caused, but not for sending the letter—I wanted Francis to realize that he could not continue to make
policy for the human genome project as if the Sanger Centre did not
exist. Several outcomes followed in quick succession: a somewhat
frosty conference call with the rest of the genome center leaders, and
then an invitation to meet the head of the NIH, Harold Varmus, in
London, and to go to Washington a month afterwards to address the
funding agency's Scientific Advisory Council. The letter was clearly
being taken seriously.

The London meeting, hastily fitted in to Varmus's already busy
schedule, took place over breakfast in a café in Soho, horribly early
on a dismal November morning. Martin Bobrow accompanied
me from Cambridge and Mike Dexter, who had recently taken over

from Bridget Ogilvie as director of the Wellcome Trust, also joined us. I argued that the Sanger Centre and Bob Waterston's lab in St. Louis were ready and able to produce mapped clones for sequencing, that currently only our two labs and Celera had put in place the high level of industrial organization needed to accelerate the production of sequence, that it would be unwise to try to implement a major change of strategy under the sort of time pressure Celera's appearance had imposed, and that the Sanger Centre was committed to finishing its chromosomes. I didn't feel at the time that we made much headway. Varmus seemed unwilling to make a complementary commitment on behalf of the United States labs to finish the chromosomes not spoken for by non-United States labs (although he later explained that he could not do this without consulting Francis). He queried the fact that I had not included the Whitehead Institute in my list of centers that were ready for a massive scale-up, which confirmed what I had begun to suspect: that despite the relatively modest contribution (compared with St. Louis and the Sanger Centre) he had made to date, Eric Lander had consolidated his position in the eyes of the NIH hierarchy as a leading figure, perhaps the leading figure, in human sequencing.

I wasn't sure about going to the Advisory Council meeting—I felt that I was being summoned to give an account of myself. It was Jim Watson, who had been very active behind the scenes in opposing the random clone strategy and retrospectively approved of the 'friendly fire' e-mail, who sowed the doubt in my mind: 'Don't go to Washington!' was his advice. But as the date in early December drew nearer, it seemed there was more to be gained by going than not. The message that high-level international consultation was important seemed finally to be getting through, and a meeting of the heads of the five biggest sequencing operations—Richard Gibbs from Baylor College of Medicine in Houston, Bob Waterston from Washington University in St. Louis, Elbert Branscomb from the Department of Energy's Joint Genome Institute in Walnut Creek,

California, Eric Lander from the Whitehead Institute in Cambridge, Massachusetts, and me from the Sanger Centre—was scheduled for the same visit. The fact that Bob Waterston would be there reassured me that I would not lack support. Michael Morgan was also invited to talk to Francis Collins.

The meeting of the five labs started a tradition of regular communication among what was first jokingly called the 'security council' of the Human Genome Project, and later became the G5. I was determined to do whatever was necessary to keep the international collaboration going, without compromising the Sanger Centre's position on the allocation of chromosomes. Bob and I offered to solve the problem of the supply of mapped clones by acting as a resource for the whole community, providing clones to other labs for sequencing. Eric accepted the offer, with the proviso that if the supply of mapped clones could not keep up with the pace of sequencing he should be free to choose BACs at random. (In the event, Bob's lab took over the clone supply for the United States, and eventually, under Marco Marra and John McPherson, established a central map to coordinate the entire project, which saved it from chaos. I was unable to persuade the Sanger Centre mappers to give Eric many clones at all, which was embarrassing but not too surprising, especially as the Wellcome Trust had always discouraged us from mapping for others.) The American funding agencies finally committed themselves to ensuring that the rest of the chromosomes were finished. All in all, it was a highly productive meeting.

The next day's presentation to the Scientific Advisory Council of the NIH was a formality by comparison, though politically important. I felt uncharacteristically nervous in the big room with its central table and the staffers arrayed in ranks around it. I had been summoned and had obeyed, just as a century earlier colonial leaders had been summoned to London and had obeyed. There were no tapestries or banners, indeed no decoration at all in the utilitarian government-issue room, but this was a theatre of power in the greatest empire on earth today. The NIH had an annual budget of $15 billion, twenty times that of the Wellcome Trust: why should they listen to me?

The Council was meeting to ratify Francis's plans for distributing the funds awarded by Congress for the genome project. Francis had presented the strategy for completing a draft of the genome by 2001 and finishing by 2003, as announced in October. My role was to stand up and say that Britain, in the form of the Sanger Centre, would support this strategy. Simple enough to do; but a lot was at stake. This was also the first occasion since the Celera announcement on which I had met Craig Venter face to face—in fact, I don't think I'd seen him since the 1996 Bermuda meeting. It wasn't that I had particularly been avoiding him, just that we had had no reason to meet, and I had certainly seen no point in trying to establish any kind of dialogue.

There had been a phone call soon after the Cold Spring Harbor meeting, though, a conference call out of the blue from Craig and Mike Hunkapiller to sound me out about their announcement. 'This isn't an in-your-face thing,' they said: they really wanted to get the sequence out for everyone. Indeed, would I be interested in joining them? I hesitated for a moment; nothing is absolute, after all, one must consider every case on its individual merits. How tempted Francis and Harold must have been—to avoid the conflict, make the compromise, get the job done without a fuss and live peacefully. And then I remembered the track record: the attempt of ABI to control assembly software, the human ESTs for sale, the forcing out of Jim. Wherever these two people had been, the decisions came down to profit. I was back in the Claremont tower. 'No, thanks,' I said. We talked for a little longer about their plan to build the best possible database, and a thought occurred to me, not for the first time. 'You're trying to be a Microsoft of biology, aren't you?' I said. 'Oh no,' was the reply, 'it's not like that at all.' A little while later, I started to see articles comparing Craig Venter with Bill Gates. Some might see this as a good thing, but the genome is a unique resource, not a technology.

Craig sat down next to me at the Advisory Council meeting; I greeted him and shook his hand, but received no sign of friendliness in return. It was very strange to be sitting next to the man I had come to regard as the potential destroyer of all that we had worked for.

Craig was there partly because of another development, even more serious than the random clone business, that threatened the fragile alliance among the publicly funded sequencing labs. Unknown to Francis or any of the groups funded through his institute, Ari Patrinos at the Department of Energy had drafted a memorandum of understanding with Celera to provide Craig with BAC clones in return for collaboration on the sequencing of the Joint Genome Institute's human chromosomes 5, 16, and 19 and equivalent regions in the mouse, with the specific aim of looking for gene families that might make important drug targets. Francis had got to hear about this, and had let the rest of us know, only a couple of weeks previously. Clearly the Department of Energy was under political pressure to collaborate with the private sector; but the terms of the draft memorandum with Celera contravened the Bermuda agreement on free release of data to which the department's scientists had signed up.

Under pressure from the rest of the group leaders at the 'security council' meeting the day before, led by Eric (who gave a magnificent and passionate denunciation), the group funded by the Department of Energy withdrew from negotiations with Celera. So it was not surprising that Craig did not look best pleased when he appeared at the Advisory Council meeting, particularly as he had been told he could not speak. At the last minute he was invited to comment on reports of a change in Celera's business plan. He complained that he had nothing prepared, but said that nothing had changed since his previous testimony, and that if there were to be a change it would be announced. He declared once more, as he had at the congressional subcommittee in June, that he had no intention of knocking the public effort, and earnestly wished to work alongside it. But he seemed unable, or unwilling, to appreciate that his commercial ambitions would be in conflict with the HGP policy on data release and therefore an insuperable stumbling block.

I returned to the U.K. feeling better about the whole thing than I had for a long time. Maybe the DOE–Celera episode had made everyone realize the absolute necessity of the publicly funded teams working together. To have seen off that threat, potentially found a way round supplying mapped clones to sequencing centers and presented a show of unity to the great and good of the NIH was, all in all, a good couple of days' work for a project that had looked in danger of fragmentation on more than one occasion in the previous months.

And I had a celebration to look forward to. Nine and a half years after the moment when the prison door clanged shut at Syosset, the worm sequencing consortium was to publish the sequence: the first of an animal, and indeed the first of any multi-celled organism. It came out in *Science* during the second week in December, and the NIH and MRC held press conferences on each side of the Atlantic to mark the occasion. Bob Horvitz remembers the event vividly—he was at the Washington press conference.

> On stage we had Francis Collins, Bruce Alberts from the National Academy of Sciences saying what a great thing this was, Harold Varmus, Bob Waterston and me. And then sitting on a little pedestal was a TV monitor, and on the screen, full face, was John.

In his talk, Bob Horvitz said the achievement was 'better than landing on the moon': both were human achievements, but the worm genome sequence was an achievement you could use for science and medicine. On our side we went down to the Royal Society with a group of people from the lab. I talked about the sequence and Jonathan Hodgkin, my contemporary from the early days with

Sydney who still worked on worms at the LMB, talked about its implications for worm biology.

The truth is that getting the worm done had been an underlying objective for me all the way through. Of course I had my day job to do, directing the Sanger Centre, and I was always there when the HGP needed me; but all through the alarms and excursions of 1998—the doubling of our goals for human sequencing, Celera, friendly fire—I was spending time in the lab helping to close gaps in the worm sequence. And between us Bob Waterston and I and everyone else involved made sure that the worm got done. That was our own: if you don't do your own project you have nothing. If you can make something else happen as well, then that's great, so I don't regret becoming involved with the human genome. But I definitely had a sense, at the end of that year, of having achieved what I had set out to achieve. Even then, the worm sequence still had some tricky gaps in it, and I looked forward to working on some of the remaining difficulties during the coming year.

The worm had been successful in a larger way, too: it had been one of the sources, right from the Room 6024 days in the early 1980s, of a new field of biology. That little trickle had become a stream, and had joined other streams to form the river that is called genomics and is transforming the way people do biology—not by displacing real experimental work, but by empowering it with new tools and entry points, with access to the totality of the codes of life. It was all working, and so for me, just like the time at the end of the cell lineage, it would soon be over.

This was different from the last time, though. The scale was vastly larger, for one thing. But more importantly, and partly driven by genomics, biology had undergone an economic sea change—it now held the promise not only of tremendous knowledge and great benefits to humankind but also of fabulous wealth. As biologists we had lost our innocence; we were out in the so-called real world and were reacting to it in a variety of ways. Before, I hadn't had to face up to the true extent of venality in the world, because it hadn't impinged on me directly. Now it began to look as though I wouldn't be able to do science any more, but would have to join some ill-defined political struggle, to try to make some contribution, however small, to turn the tide.

In practical terms I had a sense of having discharged my duty to the lab and the Trust by having at last secured the funding for a third of the human genome. This was the guarantee that the genome would remain international and that the Trust would come out with honor, having secured its own primary objective in funding the Sanger Centre. Jane was firmly in charge of sequencing (as well as being overall project manager for our part of the Human Genome Project) and it was going better and better all the time: I no longer went to many meetings, feeling more useful out of the way in the lab. And so it seemed the right moment to give in my notice as director.

It wasn't a spur-of-the-moment decision. I'd been quite open with Richard, Jane, David and my other close colleagues about wanting to step aside as soon as things were running smoothly. At the end of August I had written to the BoM as well as the directors of Genome Research Ltd, telling them in confidence that I wished to make an announcement by the end of the year, and aimed to be replaced by the end of 1999. These things take time, and Alan Munro warned me that two years would probably be more realistic, so it was time to start the process. I reminded them that I had always been a reluctant director, and that my particular interests meant that I was not going to be part of the post-sequencing phase of biology. 'In the changing environment of biology today,' I added, 'my instinctive reactions are becoming less and less appropriate.' What I meant was that I foresaw a time when centers like ours would move on from producing genome sequences for the benefit of all to a more targeted approach to understanding genes and their products. It seemed inevitable that such developments would involve questions of intellectual property and commercial development which had never formed part of my thinking.

I was confident that with the distributed system of management we had in place, the Sanger Centre was in robust shape and would not be in any way damaged by my going. Indeed, a change of head would be helpful; the genomics field was moving so fast that the lab needed to be renewing itself by 2000 if it was to retain credibility. But equally it was obvious from the reactions to my confidential letter that unless I forced the issue the Wellcome Trust would keep jolly-ing me along and not do anything about finding a successor. They issued their announcement in the middle of December. I had been right to get things moving early—my successor did not take office until almost two years later.

One thing was clear: I had no intention whatever of becoming a public face of the human genome.

6 PLAYING POLITICS

SO IT WAS ESPECIALLY ODD TO FIND MYSELF, ON JUNE 26, 2000, BEING treated as a minor celebrity. Even before I'd given my talk the photographers were crawling up and saying, 'Psst, John! This way!' and I had a little taste of what it's like being in the public eye. After the talk I was just surrounded by lenses all pointing at me, and I had to be dragged away because interviews were lined up. I thought, 'How strange—they've cottoned on to this, they do believe it's important.'

It was the day on which joint press announcements were made in London and Washington that the draft human genome sequence had been completed by both the publicly funded Human Genome Project and by Celera Genomics. I had gone with my Sanger Centre colleagues, dressed in unaccustomed suits, to face the assembled press in the lecture theatre of the Wellcome Trust headquarters in Euston Road. It was packed. The director, Mike Dexter, said a few words, followed by the science minister Lord Sainsbury, Michael Morgan, and members of the lab. In the afternoon some of us went off to 10 Downing Street for a Bill and Tony show over a video link with the press conference that was going on in Washington. We were a bit outraged by Tony Blair's speech, which we had no hand in, because amazingly it mentioned Celera but not the Sanger Centre. But the deal was that we were going to be cheerful and nice—that was the official Wellcome Trust line.

My main aim was to say to the journalists that I thought the people at the Sanger Centre had done a wonderful thing for humanity, and I thought that they ought to be supported. To be fair, we did get some halfway reasonable coverage afterwards. I had been incensed by the lack of appreciation that had been shown until then. Everyone had focused on the 'race'—were our methods better than Celera's, who's going to get there first?—and hardly anyone was saying, 'Hang on a minute, there's one bunch of people who are actually doing this for the benefit of humankind and another bunch who are trying to do it for their personal gain.' And since then there have been more articles drawing that distinction.

Moon landings excepted, it is almost unprecedented for any head of state, let alone two at once, to identify themselves so closely with a scientific advance. The Human Genome Project is politically sensitive on at least three counts: it is perceived as expensive (though not by comparison with moon landings); access to the information is of immense commercial value; and there is widespread public concern about the way that information might ultimately be used. For these reasons all of us in the field are aware that the direction of our science to some extent depends on attitudes in Downing Street and the White House.

As the champion of private enterprise, Craig Venter had lost no time in attempting to get the United States Congress on his side. All through that period between 1998 and 2000 it was very important that the British-based Wellcome Trust was so influential. If the HGP had been just an internal United States matter, then, with enough support for Celera in Congress, I feared that the role and views of the National Institutes of Health might have been suppressed.

It was only gradually that the political pressures on the G5 became apparent. To begin with, we were much more concerned with setting our scientific goals and getting into place everything that was needed to reach them. In February 1999 the G5 met at the Baylor College of Medicine in Houston to plan its strategy for producing a draft sequence. I went with Michael Morgan to represent the Sanger Centre. On the agenda was a radical plan to move forward even faster than had already been announced in September: Francis was now talking about producing a 'working draft' sequence within little more than a year.

> That was a turning point meeting. I had come to the conclusion that this was what we had to do about a week before, and hadn't even told my own staff what I was going to propose. I made the case that we should focus on getting draft coverage, and based on the capacity that I thought this group of five had, that we might be able to do that in a year.

Francis wasn't sure how his proposal would be received. He was pretty sure that Eric Lander would be in favor of it, as he had previously argued along the same lines, and Richard Gibbs at Baylor would probably go for it too. But he was much less sure about Bob Waterston or the Department of Energy genome center, and he didn't know where I would come down.

> It was pretty clear that John's opinion was going to be definitive. If he was against it, especially as he is somewhat uncompromising, it was going to be difficult to carry the day.

The discussion went back and forth for most of the day. Bob, who wasn't at the meeting but was in touch by conference call, was really not convinced that what Francis proposed was feasible. He was afraid that we might just end up by shooting ourselves in the foot by deviating from the pathway that had already produced about 15 percent of the sequence in finished form, and which could get us about half of it by the end of 2000 if we stuck to it. Much of the effort to accelerate the mapping part was going to fall on his head, and producing all the mapped clones needed to go into the sequencing pipeline was going to be a huge job for him to take on.

I knew that many of my Sanger Centre colleagues would agree with Bob. But I said I thought we should go for it. I was no less committed than Bob to getting fully finished sequence eventually, but on the other hand I'd always been in favor of pushing out unfinished sequence as fast as possible. It is useful to the biological community—the worm project had proved that—and, more importantly in the light of the Celera threat, it put the sequence itself, though not possible future uses of it, beyond the reach of patents.

In Francis's words, that 'turned the tide of the conversation.'

Bob came around, and by the end of the day John was up at the blackboard dividing up the chromosomes, and we had a strategy—it was pretty much clear from that point on what the pathway was going to be.

The accelerated timetable was announced in mid-March, and succeeded in taking almost everyone by surprise. For once the HGP caught Celera on the back foot; the company's target completion date at this stage was still 2001. Reporting the development in the *New York Times*, Nicholas Wade wrote, 'If met, the new date set by the consortium could allow the public venture to claim some measure of victory over its commercial rival, the Celera Corporation of Rockville, Md.' Craig, uncharacteristically lost for a pithy rejoinder, said that the new timetable was 'nothing to do with reality', and that it was just 'projected cost, projected timetables.' The irony of that remark, coming as it did from someone who had almost managed to scupper the entire publicly funded project with a press announcement based entirely on the projected performance of machines he had not yet even bought, was evidently lost on the writer.

Just as surprised as Craig were five of the United States genome centers that had participated in the pilot programs, but that now seemed to be left out. The NIH's announcement included the award of grants worth $81.6 million in total over ten months to just three centers: Bob's lab at Washington University in St. Louis, Eric Lander's at the Whitehead Institute at MIT and Richard Gibbs's at the Baylor College of Medicine in Houston. Glen Evans of the University of Texas Southwestern Medical Center probably made the understatement of the year when he told *Science* magazine, 'It's kind of upsetting for all of us.' Francis assured them that there would be another funding round soon that would include the smaller centers, but there was no getting round the fact that, with the funding plan he had presented the previous autumn, he had deliberately created a two-tier structure within the NIH program that greatly favored some at the expense of others.

The internal dynamics of the consortium were now completely altered. Until the launch of Celera, St. Louis and the Sanger Centre had been the biggest sequencers in the world. We each thrived on friendly rivalry with the other, but in all important matters—the regional approach to selecting clones for sequencing, free release of data—we saw absolutely eye to eye. We had no need to set each other targets, because every individual in each of the two labs was eager to stay just a little bit ahead of the other. But there were no secrets between us, and if we found a better way of doing something, we shared it. We also made a point of encouraging other international centers, for example in France, Germany and Japan, that were having a much tougher time than we had in persuading their governments to fund sequencing projects that adhered to the Bermuda Principles.

Now things were very different. Bob no longer held such a pre-eminent position in the United States sequencing community. Ever since we had launched our unsuccessful bid to go for a draft sequence in 1994–5, the two of us had anticipated that the St. Louis–Sanger axis would lead the charge when the moment finally came. We had thought in terms of a third of the genome for the Sanger Centre, a third for the Genome Sequencing Center in St. Louis and a third for everybody else. But it gradually emerged through 1998 that Bob was not going to get enough of the genome project funding to sequence a third of the genome. When Francis announced the share-out of the funds he had won from Congress to support the accelerated draft strategy until 2000, it was Eric Lander who came out on top. He was awarded $34.9 million to Bob's $33.3 million, with Richard Gibbs at Baylor College of Medicine taking $13.4 million. (At the same time the Department of Energy put $40 million into sequencing at its Joint Genome Institute.) But Bob was not surprised at the outcome.

> I got all that I had requested. I had basically failed to persuade our group that we should aim higher, and there was a sense of wanting to 'play fair.' Also our building had begun to run out of space by that time. We had missed our moment.

Eric was proposing to spend almost all of his money on shotgun sequencing, while the Sanger Centre and St. Louis retained their commitments to map, sequence and maintain a certain level of finishing—in addition, in Bob's case, to providing mapped clones for Eric and everyone else. Everything was now in place for the Whitehead genome center to become the biggest, measured strictly in terms of raw sequence output.

Now, too, Bob was seriously ill. The year had begun with the shocking news that he had cancer of the bowel. Although his doctors thought his chances of survival after chemotherapy and surgery would be at least 80 percent, I had to face the prospect that I might lose my closest colleague and a very dear friend. Bob was incredibly courageous, and never stopped working; he couldn't travel, but he joined in the Houston meeting by conference call and continued to participate in almost all of the regular Friday conference calls that had begun to take place among the G5 to coordinate activities. He went in for surgery in April, and even while convalescing surrounded himself with a mini-office so that he could keep in touch. As so often before, Bob dwarfed everyone with his sheer capacity to cope. 'He's a soldier,' Richard Gibbs commented to me. To our enormous relief his treatment was a complete success and by the middle of the year we knew he was out of immediate danger. Bob was and still is a very fit man, regularly cycling from his home to his lab—which may be normal in Cambridge but is regarded as decidedly eccentric in St. Louis.

Like the smaller United States labs, the other international centers were completely taken aback by the announcement that the target date for a working draft was now spring 2000. The very existence of the G5 was a slap in the face to colleagues who had participated in the Bermuda meetings since 1996 and regarded themselves as partners in the consortium. André Rosenthal, who headed the Institute of Molecular Biology in Jena, Germany, was particularly bitter about it. 'The policy was not agreed upon in the same international spirit as had [been cultivated] in the past,' he told *Science* magazine. 'This announcement gives the impression that [we're] not needed.' André's bitterness was understandable. Under pressure from the British and American scientists, he had put his career on the line in fighting the German government to agree to the policy of free data release, and was expecting about 7 percent of the sequence—including a large part of chromosome 8—to be produced at his institute. Now it seemed that his defense of international co-operation was being rewarded with an almost complete takeover of human sequencing by the large and well-funded United States and British centers. Japan was in much the same situation, with Yoshiyuki Sakaki of the Human Genome Center at the University of Tokyo close to completing part of chromosome 21 and already stocking up with capillary sequencers ready to make a substantial contribution to the draft. Jean Weissenbach in Paris had so far done little human sequencing, but had done important work on other genomes and on human mapping.

With the Sanger Centre as the only non-United States member of the G5, Michael Morgan and I felt a particular responsibility to represent the interests of the rest of the international genome sequencing community. Looking back through my correspondence, 'What about the rest of the international partners?' seems to be a constant refrain. I thought it was important that they remained part of the project; but at the same time I knew they would have to keep up. They all finally got together at the 1999 international strategy meeting, which again took place at Cold Spring Harbor in May. Time was tight, and it seemed a good idea to tack the strategy meeting on to the annual genome sequencing symposium as most of the participants would be there anyway. I didn't go, but Jane, Richard and David reported that the meeting began with everyone in a state of high anxiety. Francis offered apologies all round—but he then said bluntly that 1999 was going to be a make-or-break year. There was no doubt that Celera would announce that the genome was 'complete' within little more than a year, and that Congress was under real pressure from some quarters to shut down the publicly funded genome effort. The only hope was to move fast and in a tightly coordinated way, putting most of the resources available into a small number of big centers.

What the French, German and Japanese groups needed more than anything was a statement from the project's leaders that it remained an international enterprise in which they had a role. By the end of the meeting they had won recognition of their claims on certain parts of certain chromosomes, but only as long as they could keep up with the rest. There would be no question of leaving whole chromosomes to groups that did not have the funding and resources to contribute working draft sequence by the spring 2000 deadline. They were on board, but there was no doubt at all who was in the driving seat.

The allies in the publicly funded project were tightening their organization to turn the tide against the threat from the invader. It would be easy to see this as an overreaction to the entry into the field of one competitor, as against the combined forces of the international consortium. And indeed, Celera's PR operation was fond of presenting the company as a David up against the federal Goliath, Craig Venter the maverick entrepreneur against the mighty National Institutes of Health establishment. The political riff plays well to this day, and is much beloved of some British reporters, as well as a majority of the United States ones. The truth, of course, is very different: Celera was much more powerful than it appeared. Representing private enterprise, the company could count on the backing of many in Congress who were philosophically opposed to state-funded projects. By repeatedly hinting that the government was wasting its money, Craig clearly aimed to influence congressional policy on the funding of the Human Genome Project, if possible to the extent of shutting it down. We could not allow that to happen.

The Sanger Centre had managed to avoid being left out in the cold with the rest of the non-United States groups simply by being big—at that time the biggest, in terms of sequence output. In June 1998 we had held a party for everyone at the center to celebrate passing 100 megabases of finished sequence (including all species), a landmark which we were the first to achieve. (The event had bouncy castles and loads of drinks and nibbles, and the tradition continues: in 2002 we marked the passing of the first gigabase—1000 megabases.) But the new draft deadline was still going to be a huge challenge. We had to treble our output of sequence in less than a year, and unlike Eric we intended to keep our mapping and finishing activities going at the same level as before. I had really stuck my neck out to defend our claim to one-third of the genome. Now we would have to deliver under the close scrutiny of all our partners in the international consortium. It was all terrifyingly public. With assembled sequences of anything more than 1 kilobase being added to the public database every day, anyone could see what we had—or had not—done. In addition, the regular Friday G5 conference calls required each center to report on its production for the past week and make a prediction for what it would produce in the weeks to come. This was crucial to keeping everything on track, and the office of the National Human Genome Research Institute did a great job of book-keeping.

It was a very different way of doing science. Most projects take as long as they take, although scientists are usually in so much of a hurry to get results that there's no need to crack the whip. But we had voluntarily put ourselves under tremendous time pressure. Not every member of the G5 cared about beating Celera, but as long as some of them did, the rest had to run at the same pace or risk losing sequence to faster-moving centers. And ramping up production could not be done overnight.

We needed not only new machines, but a new type of machine. Of course, all the sequencing centers now wanted to buy capillary sequencing machines of the type that Celera was installing. Perkin-Elmer had been very clever in launching Celera. At the time many argued that to go into competition with your customers was a questionable business move. But in fact Perkin-Elmer would win out whatever happened. If Celera put the HGP out of business, it would earn the financial rewards of monopoly ownership of the genome sequence. If, as it turned out, the HGP decided it wanted to match or exceed Celera's effort, then it would have to go to ABI for the new capillary sequencers—at $300,000 a time—in order to do it. In other words, by launching Celera, Perkin-Elmer had hugely increased its market for the 3700 machine, and for the expensive reagents with which it had to be fed. Tony White, chief executive of Perkin-Elmer, knew exactly what he was doing. He was later quoted in *Forbes* magazine as saying, 'The day after we announced Celera, we set off an arms race . . . Everyone, including the government, had to retool, and that meant buying our equipment.' ABI—which formally changed its name to PE Biosystems in the spring of 1999—could scarcely keep up with the demand, and Mike Hunkapiller found himself the target of complaints from each side that he was favoring the other. But I'm sure he found this a minor irritant set against the sales of over $1 billion that PE Biosystems reported in the year following the Celera launch.

As long-standing customers of ABI/PE Biosystems, we had known about the new capillary technology for some time. Indeed, it was in part the demand from the publicly funded centers for an alternative to slab gels that had driven this development. Nor was PE the only company making capillary sequencers, although after evaluating the others we (and most of the other genome centers) decided to go for the 3700. The change would have knock-on effects throughout the whole enterprise: the new machines used different chemistries to carry out the sequencing reactions, and needed different organizational procedures for handling the samples. We also needed more space to put them in, as well as building new, automated systems for picking clones and preparing the templates for sequencing. In theory, space was not a problem at the Sanger Centre. We had not yet occupied a wing of the building, called the West Pavilion, which had been built as a shell against the need for future expansion. But it needed to be fitted out with labs, services and equipment, all of which would take time.

There were challenges all through the lab. The mappers had to increase the clone supply; more subclones had to be made, more samples prepared, more data handled. The burden of implementing the sequencing scale-up fell on Jane Rogers and on Stephan Beck, head of human sequencing. The clattering of the clogs that Stephan wears around the lab became more noticeable than usual. Jane hoped to limit the extent of our change-over, to limit the costs, and she initially ordered 30 of the new capillary machines. Eric ordered 125, further confirming his ambition to become the biggest center. We still had something like 140 of the old 377s, many of them now adapted to run 96 lanes per gel. We were running them three times a day, so that was almost 450 slab gels a day to prepare. That was the part that wasn't easily scalable, and the reason why the new technology was attractive. At least we were able to keep a solid base of production going while the new capillary machines were being brought in—although with hindsight we could probably have gone faster in the end if we had replaced them. For much of 1999 the attempt to keep up the pace was a nightmare. Even the computer system crashed repeatedly for nearly a month in July and August. Phil Butcher, the Sanger Centre systems manager, and his group had a terrible time sitting up at nights waiting for the elusive hardware fault to recur. He eventually tracked down and replaced the offending hardware and after that everything was fine. One way or another we fell behind on our production targets, and our United States colleagues let us know that they were pretty unimpressed. Once things were working again, however, we rapidly increased our output.

As well as the need to meet our targets for the working draft, we had taken on another major responsibility. In April a consortium consisting of the Wellcome Trust and ten of the big pharmaceutical companies had launched a project to mine the sequence for variability between individuals. Each person has two copies of the genome: one from each parent. Between any two copies 99.9 percent of the sequence is identical—which is why we belong, recognizably, to the same species and can have children together. But the final 0.1 percent, roughly one nucleotide base in every thousand of the 3 billion total, differs from one copy to another. At a particular point two-thirds of the copies might have an A, for example, while the remaining third have a T. These differences are called single nucleotide polymorphisms or SNPs (pronounced snips). They are the exceptions in the human genome sequence that make us individuals rather than identical clones. And although we are so similar to every other human being on the planet, that still leaves us with millions of possible differences between ourselves and someone else. There are other sorts of differences, such as deletions and longer replacements, but SNPs are the most common.

In practice most SNPs probably have no effect at all, because they do not fall in protein-coding or regulatory regions of the genome. But others give us brown eyes rather than blue, affect our height, or influence the degree to which we are creative or impulsive. And some have significant effects on our health. They do not necessarily make genes defective in the same way as the rare mutations that cause muscular dystrophy or hemophilia, but they could cause subtle differences that influence our susceptibility to heart disease, for example, or how well we respond to a particular drug. And mostly it will prove to be combinations of SNPs, rather than individual SNPs in isolation, that cause these subtle effects.

SNPs are of enormous interest to drug companies, who want to use SNP maps for a variety of purposes, including the development of drugs tailor-made for a subset of patients according to their genetic profiles, and in the longer term the identification of new drug targets. Just as with ESTs and then with the complete human sequence, there was a fear that the gold-rush mentality would lead to large numbers of SNPs being tied up in patents and pulled beyond the reach of further research. That fear seemed to be justified in May 1998, when Celera announced that a 'catalogue of human variation' (that is, a SNP database) would be one of its flagship products. Other commercial operations, such as Incyte and Curagen, almost immediately countered by launching their own SNP initiatives, focusing only on the protein-coding regions of the genome.

Realizing that for each company to build up its own SNP data-base would be enormously wasteful of time and resources, Glaxo Wellcome plc began to talk to a number of other big pharmaceutical companies about doing it jointly. Michael Morgan got wind of this, and offered the resources of the Wellcome Trust—money and the Sanger Centre's sequencers—to help get the project off the ground. Despite the difficulties of getting ten competing companies to work together, and the need to avoid falling foul of the stringent anti-trust laws in the United States, the non-profit-making SNP Consortium was formally launched in April 1999, with a budget of $14 million from the Trust and $3 million from each of the ten companies. Alan Williamson, now retired from Merck, played a key role in the negotiations, in a reprise of his earlier success in brokering the fund-ing of Washington University to carry out EST sequencing in 1994. The SNP Consortium commissioned the Sanger Centre, the Whitehead Institute and the Genome Sequencing Center in St. Louis to find 300,000 SNPs by 2001.

The SNP Consortium database was to be free and publicly accessible; in the jargon of the commercial world, it was to be seen as a 'pre-competitive' development and therefore not bound by the laws against collusion by companies in the same industry. Its announcement struck a timely blow for the common ownership of the genome and caused a blip in the share prices of the genomics companies. This public ethos meant that for the Sanger Centre to work for the consortium would not conflict with the principles we had established for sequencing, and the contract brought in valuable funds. In practice we had already started the work, as poly-morphisms frequently turn up in the overlaps between one length of DNA and another as the sequence is assembled. Ian Dunham and his colleagues had also begun to look for SNPs more systematically, examining the overlaps in stretches of sequence on chromosomes 22, 13 and 6. But the consortium contract now obliged us to work more comprehensively. The pilot work went well, but our problems with getting the new sequencing technology working meant that by the autumn we were seriously falling down on our targets and under threat of having our contract terminated. David Bentley averted the threat by honestly setting out our problems at the Sequencing Consortium meeting in October, and confidently asserting that we would catch up by the end of the year.

If we seemed to be having problems achieving our goals, it was only because we were trying to do so much. Throughout all the setbacks we were making steady progress with our human sequencing projects, and knew that we were building up to a major triumph. In August 1999 Ian Dunham, who coordinated the chromosome 22 sequencing project, sent round an e-mail telling everyone involved that the team now had a contiguous sequence of DNA 9 megabases long—5 megabases longer than any other human sequence in the world. They were well on target to publish the complete, finished sequence by the end of the year.

Chromosome 22 is one of the smallest chromosomes, containing less than 2 percent of total human DNA. The chromosome 22 sequencing project was an excellent example of the Human Genome Project in miniature. The Sanger Centre carried out over two-thirds of the sequencing, in collaboration with Bob Waterston's lab in St. Louis, Nobuyoshi Shimizu at Keio University in Tokyo and Bruce Roe at the University of Oklahoma. Five other institutions in the United States, Canada and Sweden had worked on the mapping phase. During 1999, as we struggled to produce shotgun sequence for the 'working draft' of the whole genome, Ian and his colleagues were patiently working through the time-consuming finishing stage on chromosome 22. They had to link up all the sequenced clones, correct errors and look for gaps. Filling the gaps meant trying to find new clones that were clearly linked to landmarks on either side of the gap and sequencing those. With our map-based methods it was possible to do this in a systematic way, and Ian's group and the sequencing teams closed gaps remorselessly for month after month.

The chromosome 22 sequence was published in *Nature* on 2 December. It would always have been a major milestone, but with the publicly funded project under such pressure from the Celera PR machine, we had to make as much of it as we could. Just as we had with the worm, we held simultaneous press conferences in Washington and London, and this time there was also one in Tokyo to mark the important Japanese contribution. We've since been mocked for some of our hyperbole on that occasion. In newspaper interviews I made comparisons with Copernicus's discovery that the earth goes round the sun, or Darwin's theory that humans are relatives of apes, and Mike Dexter said something similar about the invention of the wheel. But of course I wasn't just talking about chromosome 22—I was thinking about the whole enterprise of molecular biology, and how it is changing our view of ourselves. I've used the same comparison frequently, and I don't think it's over-stating the case.

What was immediately important to us about the finishing of chromosome 22 was that it proved that the strategy we had adopted for the whole genome worked. Ian Dunham had his feet more firmly on the ground when he told *Nature* that the main significance of the publication was that 'it shows that you can get very good finishing using the clone-by-clone approach.' If we could do it for chromosome 22, we could do it for the whole genome, and suddenly that long-cherished goal seemed a lot closer.

Despite its small size, chromosome 22 had plenty of significance in its own right. Geneticists had already implicated it in at least thirty-five diseases, including schizophrenia, chronic myeloid leukemia and some forms of heart disease. Using the unfinished sequence that had been released over the previous few years, they had begun to pinpoint some of the genes involved. James Scott, professor of molecular medicine at the Hammersmith Hospital in

London, identified seven new genes relevant to cardiovascular disease in this way. 'We could not have done this work without the chromosome 22 data,' he told *Nature*. The publication of the complete sequence not only announced that almost 33.5 megabases of finished sequence was in the databases, but included the identification of 545 genes (based on comparisons with other gene sequences such as ESTs), more than half of which were previously unknown in humans.

The other important point about the chromosome 22 publication is that it brought home to us and to the scientific community at large what a very difficult business it was going to be to sequence the complete human genome. The published 'complete' sequence in fact ignored completely the short arm of the chromosome, and included eleven gaps ranging from 50 to 150,000 base pairs on the long arm. The short arm consists almost entirely of repeat sequences which make it all but impossible to reassemble it in the right order using any current technology. But its composition makes it very unlikely that it contains many protein-coding genes, if any. The 11 gaps in the sequence of the long arm were left because for reasons that we don't fully understand, those regions refused to establish stable BAC clones. Some of these gaps have since been closed, and more will be closable in time, but it would have been unnecessarily punctilious to delay the publication until the most intractable problems had been solved. The sequence as published was still a major achievement.

It was very satisfying to be able to make a statement about the progress of the publicly funded genome project that really meant something. I had been very conscious throughout 1999 of the political high wire we had to walk in relation to Celera. Craig Venter's agreement with Gerry Rubin to sequence the fly genome was announced in January 1999, including a commitment to make all the data publicly available. On the back of this, Craig immediately began to negotiate with Francis Collins about a similar agreement on the human genome. From the moment I saw the draft agreement I could see that the advantages would all be in Celera's favor. For example, while it included a general comment about commitment to making the sequence available to the international scientific community, it crucially left vague the question of free and unrestricted access. It also included avoiding 'inappropriately adversarial comments' about each other's work, just as on our side we were beginning to talk about a more vigorous press stance.

I was very averse to the whole notion of entering into an agreement with Celera. I recommended to Michael Morgan that whatever Francis might do, the Wellcome Trust should not be party to it. Together with Bob Waterston's lab in St. Louis, we had already had a direct approach from Gene Myers, one of the original proponents of a whole-genome shotgun approach to the human genome who now worked at Celera. He asked us to hand over all our trace data on *C. elegans* so that they could use them (minus any information about map locations) to test their whole-genome shotgun assembly program. It was not a trivial request—it would mean tying up someone's time extracting all the trace data and saving it in a form that could be transmitted to Celera. Craig became highly indignant when we seemed reluctant to agree, arguing that as the work was publicly funded the data should be available to anyone. Of course, we had always released our assembled sequence freely, but now it seemed that wasn't enough; he wanted the raw data, too. We continued to negotiate in a desultory way—for example, we proposed that Celera should make its shotgun assembly program available to us in exchange; they offered to pay for the extra work involved—but once they were up and running with the fly sequence they had no need of the worm data any more, and the subject was dropped.

In the United States, however, there were repeated attempts by the public funding agencies to come to some kind of compromise with Celera. There was discussion of a plan, for example, to enable the company to deposit its data in a special section of GenBank, where anyone could look at it on the web, although researchers would not be able to download it as they could with the public data. This too came to nothing. The 'public release' that Craig had promised when Celera was launched seemed less and less likely to happen, at least in any form that bore meaningful relation to our definition of public release.

Meanwhile a relentless barrage of Celera press releases made it look as though they were simply blowing the public project out of the water. They started sequencing the fruit fly *Drosophila* in May 1999, and in September announced that the sequencing was 'completed.' This did not mean that they had fully finished, or even assembled, the 180 megabase sequence: it just meant they had run enough samples through the machines to cover the whole genome. But of course the message that came over was that the fly genome had been finished in four months, and needless to say Celera lost no opportunity to make unfavorable comparisons with 'other early genomes' (presumably including the worm) that had, in the words of its press statement, taken 'over a decade' to complete. (At the time of writing, two years later, the fly genome is still being finished, as one would expect.) More seriously, the apparently happy collaboration between Gerry Rubin and Craig Venter which had produced the fly sequence in such record time was held up as an example of what the human project ought to aim for. 'It has been a win–win affair,' said *Nature* in an editorial in December.

Six weeks later Celera announced that it had sequenced 1 billion base pairs of human sequence, and the pressure was on Francis Collins again to find some way of collaborating. There is no doubt that the opportunity to add Celera's shotgun data to our mapped clones could be immensely valuable, and I had publicly supported this prospect on more than one occasion since the first Celera announcement. But not if it meant restrictions on who could use the data and on what terms. In the case of the fly, Celera had eventually agreed to deposit the data in the public databases at the time of publication, with no restrictions on its use. Gerry had been firm in keeping Celera to the agreement, telling Craig that on no account could he copyright the database so that other commercial companies couldn't use it. There was a brief panic when some of the fly researchers discovered that a notice had appeared on the website of the National Center for Biotechnology Information forbidding the redistribution of the fly data, but Mike Ashburner blew the whistle and once again Gerry stepped in to ensure that Celera stuck to the letter of its original agreement. The notice came down in less than two days. In November Celera invited forty fly biologists to an 'annotation jamboree.' Annotation is the process by which the raw sequence is analyzed for gene content and embellished with any extra information relevant to understanding its biological role. At the Celera jamboree geneticists, sequencers and bioinformatics people all got together, sitting at computer screens to add all the details they could of likely membership of gene families, comparable genes in other species and so on. The whole exercise was hugely valuable to the fly community and to biology as a whole, as many of the genes described would be relevant to other species including humans.

But the human data was a very different proposition to a commercial organization such as Celera. By late 1999 it seemed clear to me that the company was never going to agree to a joint database with completely unrestricted access. They might let people look at their data, but they weren't going to let them add to it and pass it on. In other words, as far as the human was concerned, they wanted to keep control of the annotation stage. This was simply unacceptable. The raw, unannotated genome is not a useable tool in the hands of the average biologist. What the public project aimed to provide was much more than this. Tim Hubbard, head of sequence analysis at the Sanger Centre, and Ewan

Birney, who had moved from Richard's group to the European Bioinformatics Institute next door, had been working throughout 1999 on a software tool called Ensembl that would automatically annotate the genome and display it in a user-friendly way. (It went on line for the first time in October that year, and has been regularly updated ever since, its development funded by the Wellcome Trust.) Providing an analysis of the genome was an essential part of putting it in the public domain, both to give users the best possible view of the data and to preempt trivial patenting based simply on sequence comparisons; handing responsibility for this step to Celera was out of the question.

Frankly, I thought that further negotiation would be pointless. But Eric Lander seemed to think that collaborating with Craig was the only way to avoid his being declared the clear winner in what was increasingly presented as a race to complete the genome, and we gradually realized that he had been discussing the idea with Celera's representatives for some months. Eric saw some kind of collaboration as the only way to gain control of the situation.

> I did think it would be a useful thing to try to defuse this and force the data out into the public domain. Also I thought that the wars going on during the project were very damaging. I wanted to see this as peaceful as possible, and Craig and I exchanged e-mails and conversations on all this.

We at the Sanger Centre knew nothing of this at first. Then I heard from Nicholas Wade on the *New York Times* that a proposed collaboration was being discussed, and wondered what was going on. It was not until Bob phoned me in mid-November that we realized how serious it was. He said that a conference call had been set up with Celera, that Eric had prepared a background paper, and that he thought I should be in on the call. He sent the paper to me for comment. A day or two later I got a call from Francis Collins asking me to join the conference call, which was scheduled for the next day, a Saturday. But at the last minute it was called off. Everyone was rather evasive about why—Francis just said he had decided it would be 'premature'—but it gradually emerged that Craig had refused to join in if I was on the call, and Bob had continued to insist that my presence was essential.

The following weekend was assigned for our annual board of management retreat. That year we went to Stamford in Lincolnshire. Lovely as Blakeney is, it's hard for people from London to get there, and we wanted Mike Dexter and Michael Morgan to join us for part of the time. We assembled in the George, the splendid old inn that rambles through the center of the town. I brought Eric's document and we went through it to see what we could commit to. The retreat was really for planning the future of the lab, but this issue was too important to wait. The proposal was for a merging of the data from the two sides, beginning the following spring or summer, with joint publication by the end of the year. But the document preserved the principle that the complete sequence should be freely available in the public domain, with no restrictions on its use.

Reading it, I doubted that Celera would actually sign up to such liberal conditions when it came to the point. It seemed absurd to suppose that progress could be made, but we had to go through the proposal legalistically just in case it became reality. We talked both about the minimum that we could agree to and about the safeguards that would be required to enforce the agreement if made. (I hadn't forgotten the last-minute attempt to backtrack over the fly release— and these were human data we were contemplating.) I spent the day running between the meeting room and the phone in my hotel room to confer with Bob.

There were few differences of view among us all. The Trust and the directors of Genome Research Ltd were just as concerned as the BoM about data release and freedom of use; the Trust had not invested its money to see the benefits going to a United States entre-preneur. No wonder Celera targeted the Sanger Centre and the Trust in press statements, accusing us of wasting money that should go to other kinds of research. The Trust was beyond the reach of political lobbying and so had to be attacked in other ways. Many of Craig's jibes were ludicrously wide of the mark. For example, he told the *New Yorker* that 'the Wellcome Trust is now trying to justify how, as a private charity, it gave what I think was well over a billion dollars to the Sanger Centre to do just a third of the human genome.' In fact the total Trust grant for human sequencing up to the end of 2001 was £120 million, or $180 million. It used to bother me when Craig came out with this stuff, because I knew that British scientists were hearing it and some were disliking us for the stance we took. But the shriller the accusation, the more obvious it was that we were doing something necessary and had to stand firm. And the role of the Trust in defending the public position is just as important today.

Now that we were officially in the loop on the progress of discussions with Celera, we spent the next month scrutinizing successive versions of the 'statement of principles' originally drafted by Eric and refined by Bob. A meeting with Celera representatives was finally set up for 29 December. I felt very strongly that the Wellcome Trust should be represented, to ensure that all its invest-ment in the principle and practice of free release—not to mention their investment in the Sanger Centre—should not be thrown away by an ill-thought-out agreement. For this role Mike Dexter nominated Martin Bobrow, a member of the Trust's board of governors as well as of Genome Research Ltd, and someone on whom I'd come to rely for wise advice. The rest of the public side's negotiating team consisted of Francis Collins, Harold Varmus and Bob Waterston. Celera fielded Craig Venter, Tony White, another Celera executive, Paul Gilman, and Arnold Levine of Celera's scientific advisory board.

Eric and Francis hoped that the meeting would establish common ground, and sent Celera a copy of the 'shared principles' in advance. But it was obvious to Bob from the word go that sharing was not on the agenda.

We had been led to believe that they were seriously seeking some co-operation, and that they understood that if we were to co-operate we were going to have to continue to release data. But boy, when we got in there Tony White had a different view of what was possible. He just took a hard line—is this going to make us any money?

Led by White, the Celera team demanded that the public project stop producing and releasing its own data as soon as the combined effort had sequenced the genome to sufficient depth to assemble a complete draft. A joint database would be the only way people could get access to the sequence, but Celera wanted to control it. The company could not accept the condition that the pooled data would be available to everyone, even their commercial competitors, to re-package and sell if they wanted to. Tony White wanted the merged data set to be protected from commercial use by others for three to five years; Francis Collins was prepared to consider six months to a year at the outside. 'It was so different from what we had been led to believe was the basis for us going there,' says Bob. He wondered whether the Celera team had planned some kind of nice guy/heavy routine for Craig Venter and Tony White, but 'in the end we only got the heavy.'

In the case of the fly, Celera had been prepared (at least under pressure) to put the data in the public databases. It gained a lot of credibility for accelerating Gerry Rubin's project and operating in a genuinely collaborative fashion, while giving paying subscribers to Celera a few months' advantage over the public at large in access to the data. But in the case of the human sequence, the stakes were much higher. Craig was already quite open about the fact that

Celera was going to combine the publicly available data with its own in the commercial product it produced, and the company needed no agreement from us to do this. On the other hand, Celera's representatives were very negative about the possibility that the public project might use Celera data, which at this point they were talking about releasing on DVD, to help with the finishing of the sequence. (The DVD idea was dropped altogether soon afterwards.) And they did not accept the last principle on our list: that if there were to be a scientific publication containing data from both projects, then it should have authors from both sides. The fly project had been trumpeted as a model for what could be done in the human—but Celera was clearly unwilling to conform to its own model. There was no meeting of minds; it seemed that the only reason the Celera team had agreed to the meeting at all was that they did not want to be seen to be the ones who had cut off negotiations.

There was deep frustration among the public project's scientists that with or without their agreement, Celera was going to profit from our work while simultaneously claiming to have 'beaten' us in the 'race' to the genome. As it began to look increasingly unlikely that any agreement would be signed, some of those on the public side began to wonder if there was any way we could give our data some measure of protection. Patenting had been dismissed early on in the discussions at the first Bermuda meetings, as had the idea of holding the intellectual property in some sort of trust. But our head of sequence analysis Tim Hubbard proposed a different model. I had compared Celera with Microsoft in its desire to corner the genome market, and it was in the software world that Tim found another analogy.

He was very struck by the 'free software' movement that had come up with a way of encouraging collaborative software development by ensuring that the results of the collaboration, the computer source code, would remain available to anyone in perpetuity and could not be turned into commercial property. The movement had grown from its roots in 1984 to a point where its collective software could be put together to create a complete operating system popularly known as Linux. The movement is the antithesis of Microsoft, which jealously protects its source code as a commercial secret, and it has come to be seen as a counter to the hegemony of Bill Gates's company. Anyone is free to download the software from the internet. The source code is 'open', rather than a commercial secret, and users are free to modify it and pass it on to others, either free of charge or for a fee. The only constraint is that users must agree, by signing a license, that the same conditions will apply to any modified version they pass on. (This kind of agreement is sometimes known as 'copyleft.') The result has been a spreading community of users and developers of free software, in which no-one can impose secrecy on their version or deny others the opportunity to develop it further.

As talks with Celera proved less and less likely to get anywhere, Tim and others began to work on the idea that we should use the open source model to protect our data. The idea was to put a note on the human genome data deposited in the public databases by the G5 genome centers, saying that anyone would be free to use the data in their own research or to develop products, and to redistribute it in any form. However, anyone who did this would not be allowed to put in place new restrictions on its further development or redistribution. Michael Morgan was rather taken with this idea, and the Wellcome Trust's head of legal matters, John Stewart, spent a lot of time looking at the arguments and drawing up a draft license agreement.

But the idea met with a chorus of disapproval from those at the public databases. They argued that it went entirely against the principle, hard won over the previous decades, that data deposited in the databases were completely free for anyone to use without restrictions. They pointed out that other commercial companies, such as Incyte, had for years been selling commercially protected proprietary databases that included public data and no-one had ever protested. They were vehemently opposed to encouraging the idea that anyone in future who wanted to deposit data in the public databases could impose their own set of conditions. They reminded the G5 that international collaborators who had won their countries' acceptance of the Bermuda Principles only at some cost to themselves would justifiably feel betrayed if the G5 were seen to be retreating from those principles. And finally, far from being the PR coup that Tim had envisaged, they foresaw it as a PR disaster, easily interpreted by Celera as an ill-intentioned spoiling tactic.

My own feelings were confused as the discussion swayed to and fro. Looking back, I see that Tim and I were interested in the open data possibility in the same sort of way as Francis Collins and his colleagues had been in the memorandum of understanding with Celera, as a means to escape from an absurd situation. But our scheme wanted to change the world, whereas the memorandum would have recognized the world as it then was and changed the project to fit. Still, the critics were right, of course: in the end our whole-hearted commitment to public access and free use of the data by both industry and academic scientists was our biggest selling point, and to compromise it would have been disastrous. We dropped all discussion of open source licensing or any other form of restriction on the use of data from the public project. Meanwhile, in October 1999 Celera had announced that it had applied for patents on 6,500 new genes. Although these were provisional applications, and Celera claimed it would ultimately pursue only 200–300 of them, the fact remained that they could potentially give Celera title to a great deal more biological information than it had suggested when the company was launched.

Francis Collins was more concerned than ever to dispel the persistent media image that Celera and the HGP were engaged in a race, and an acrimonious one at that. The success of Celera's clever press campaign reinforced the view that something needed to be done about it. In January 2000 the company announced that it had sequenced 81 percent of the genome, and had combined this with the public data to produce 90 percent coverage. The instant impression, to the uninitiated, was that Celera had done nine times as much as the HGP.

It's worth looking in some detail at this astute announcement. Remember that we need to sequence enough reads to cover the genome several times over in order to close most of the gaps. The Celera release was based on the fact that they had only 1.75-fold coverage in raw sequence reads at the time, which if distributed randomly would mean that 81 percent of the genome would be represented. This was purely a paper calculation, because there was no way of assembling such a thin coverage of reads to validate their distribution, but was probably about right. The extra 9 percent arose because the public project at this point had about half the genome represented in draft sequence from clones, and so on a random basis would be expected to make up half the remainder. But the two sorts of data were not comparable at all. Our side was working systematically, and the individual clones had been sequenced to a fourfold depth of coverage, which resulted in a useful level of assembly. (This could be further enhanced if it were to be combined with Celera's reads.) So an objective statement would have said that half the genome was well represented and mapped by the HGP, and a further 40 percent could be found in sparse random reads from Celera.

I tried hard to explain this to journalists, but not many got it and most thought it was sour grapes on my part, though a few more careful writers noted that half of the data in the Celera database in fact came from the public databases. The result was that the media myth of Celera being ahead of the game became firmly established, even though (or perhaps because) all our data were there for all to see and use, while nobody had seen Celera's data at all. To paraphrase Gilbert and Sullivan, 'a press release, a press release, a most ingenious press release'!

It was the first time in my life that I'd been faced directly with such ruthless manipulation. I vaguely knew that such things went on, of course, and was sadly aware that political life entailed a measure of 'economy with the truth.' But that had all gone on somewhere distant from me. Now the information was coming from a company run by a highly intelligent person, a fellow scientist, someone who claimed to be working for the interests of humanity. Was it possible that he didn't understand what he was saying? It was shocking: the methods of journalism were being used to report science. And of course the journalists loved it; these were good, clean, uncomplicated stories with none of the ifs and buts that mar real scientific reports for the purposes of the media.

At first I expected Francis Collins and Michael Morgan to deal with it, as we had agreed (I thought) beside the pool at Airlie House. If they needed professional PR support they could hire a firm to help. But they seemed unable to respond effectively. We were all thrown back on our own resources to present the case as well as we could. We suffered from lack of coordination and lack of time.

For a while it didn't bother me greatly, because I expected that once people outside knew what was going on they would rise up in protest. As the months went by, and Celera was lauded by commentators, I continued to think that it was only because nobody knew the true situation. Little by little, I found myself, through answering questions, edging into the spotlight myself, trying to explain that Celera's press releases weren't painting the true picture. And insidiously I found myself going along with the media's desire for an easily identifiable figurehead. I hadn't wanted to take a lead on the PR front for the simple reason that, judging by the usual standards in the scientific community, the human sequence was not my work. But once I started giving interviews, making my points in what I hoped were clear and forceful terms, the whole thing snowballed.

A few people paid attention, but seemingly not many. For

instance, I was sadly disappointed, and still am, by the BBC's news coverage: after all, as a publicly funded body, serving the U.K., surely it should have seen what we were doing and how important it was to keep genome data in the open? I began to realize that presentation matters enormously, that nobody has time or patience to examine the facts for themselves but rather takes up what is proffered most conveniently. So I began to adapt, to be more vehement. I still clung to the thought that if only people knew the truth they would come round to our point of view.

A significant part of Celera's press announcement was the statement that it planned to stop sequencing once it had achieved fourfold coverage—each DNA base in the genome covered by an average of four sequence reads—rather than going for the full tenfold coverage it had originally planned. The implication, clearly spelt out by *Nature*'s news team although not widely discussed, was that Celera was no longer going to rely on its whole-genome shotgun assembly program to put the sequence together. 'Celera will need to hang its sequence data on the framework produced by the public project,' said *Nature*. In other words, having initially declared that mapping was unnecessary, Celera was now preparing to use the public project's map to help assemble its own sequence.

Celera described its use of our data as a 'de facto collaboration' (I had always thought that collaborations were two-way affairs, but time moves on), but Francis was still after something more formal. After the disastrous 29 December meeting he made repeated attempts to contact Craig, to keep negotiations going at some level. If he could not achieve the merging of the data from the two efforts, at least, he thought, he could try for simultaneous publication. But for two whole months Craig became mysteriously unable to return phone calls or e-mails. The most Francis achieved was a telephone conversation with Tony White, who made it clear that he was not prepared to move from the position he presented at the meeting. Knowing that Celera was blaming the public project for the breakdown of negotiations, Francis felt it was important to tell the other side of the story. He drafted a letter setting out the main points of difference with the company and recounting his unsuccessful attempts to restart negotiations. The letter, marked 'Confidential' and signed by the four negotiators—Francis, Harold Varmus (who had resigned as head of the NIH at the end of 1999 to become president of the Memorial Sloan–Kettering Cancer Center in New York, and had been replaced as acting director by Ruth Kirschstein), Bob Waterston and Martin Bobrow—went off to Craig on 28 February 2000. It set a deadline of 6 March for his response, adding that unless they heard from him by that date, the authors would assume he was no longer interested in collaboration.

Everyone involved understood that the contents of the letter were likely to be made public at some point, although exactly how and when this should happen was left vague at first. I genuinely believed that once the world saw how intransigent Celera had been in its negotiating stance, and how determined it was to keep control of the data, everyone would immediately be on our side.

In the intervening week Celera launched its first, highly success-ful share issue, which netted the company almost $1 billion. We felt we could not release the letter until the issue closed, as it might have been seen as an act of deliberate and illegal sabotage. But we were anxious not to wait too long. The letter was released to the press at the weekend before the Monday deadline for Celera's response, and it was made known that the Wellcome Trust was the source. (It would have been politically disastrous for the National Institutes of Health to be involved in such a leak.) It had a huge impact, but not in the way we had anticipated. By jumping the gun by one day we inadvertently put ourselves in a bad light, which Celera was im-mediately able to exploit. Craig Venter and Tony White came out blazing with righteous indignation, calling the Trust's action 'slimy' and 'a low-life thing to do.' Craig even taunted us for not having released the letter earlier and dented the share issue. They told the

Washington Post that they believed the public genome project was deliberately trying to sabotage any chance of collaboration with Celera because it wanted to make a deal with another privately backed consortium in order to get the genome done first. Francis Collins had indeed been negotiating with Incyte about contract sequencing, but this had nothing to do with the breakdown of the Celera negotiations.

Amazingly, it now seems with hindsight, we had not thought at all about how we would handle the follow-up interviews, or what Celera's response might be. The United States press tracked Francis down at his family home and immediately put him on the defensive: no, NIH had not been involved in the leak, and the idea that it had been done for underhand motives was 'fanciful.' It sounded weak. Craig, for his part, issued a response couched in terms of pained dignity, insisting that he continued to be interested in pursuing 'good faith discussions towards collaboration.' He reiterated, however, his company's need for assurance that other companies would not be able to repackage and resell its data. He emerged with his credibility intact, although not his bank balance. The news that a formal collaboration was almost certainly off worried the markets. The value of stocks in all genomics companies had been climbing at a ridiculous rate since the turn of the year, along with high-tech stocks generally. Celera's own stocks increased in value from $186 to $258 on the strength of the January press release alone. Back in the autumn they had hovered around $40. In the share issue the week before the leak, PE Corporation had sold its 3.8 million shares in Celera for $225 apiece. But as soon as news of the impasse between Celera and the HGP appeared, biotech shares began to fall. (Why they should do so was a mystery to me, as with or without an agreement, Celera would still have full access to the public project's data.)

The gloves really came off in the next few days (the *Washington Post* described the genome project as 'a mud-wrestling match'). I was interviewed on the *Today* program on BBC Radio the following

Monday morning. I pointed out that our problem was that Celera not only collected their own data but would 'hoover up all of ours'—which of course was publicly available—call it their own, and charge others for using it. 'It's a sort of con-job, if that's not too rude a word,' I added. From its place deep inside the interview *BBC Online* pulled out the word 'con-job' and flashed it around the world, to be seized on by journalists. And what did people say? Some approved. But many accused me of mud-slinging, jealousy, protecting my turf. I had been heard, but the world by and large divided along party lines.

I had entered the world of politics.

A week later Bill Clinton and Tony Blair made a joint statement saying that the human genome sequence should be freely available to all researchers. The statement was the result of careful lobbying for over a year, initiated by Mike Dexter at the Wellcome Trust—it was just coincidence that it finally came out a week after the leak debacle. We were delighted with the statement, but all it really amounted to was a government-level confirmation of the Bermuda Principle that primary genomic sequence should be freely released. It had no standing in law, and it did not threaten companies' right to patent genetic sequences of proven utility—in fact it explicitly supported the protection of intellectual property. But on the day of the statement CBS Radio News reported that Clinton and Blair had agreed to 'ban patents on individual genes', following an early morning White House press briefing.

That proved to be the last straw for an already jumpy and still overvalued stock market. The Nasdaq, the index of biotechnology and other high technology stocks, suffered the second biggest fall in its history, more than 200 points. Thirty billion dollars was wiped off the value of just ten biotech companies in a day. It was left to Neal Lane, the President's scientific adviser, and Francis Collins, to put things straight to a clamoring Washington press corps at a lunchtime press briefing. The Nasdaq bounced back almost at once, but genomics companies such as Celera and Incyte stayed for some time at a more realistic level than the dizzying heights they had reached only two weeks before.

This time I was asked at the very last minute to appear on BBC TV's late evening current affairs program *Newsnight*—Don Powell, the Sanger Centre's press officer, had to come and drag me out of the pub. I was sitting in the remote studio in Cambridge, all prepared to talk about the impact of the announcement on patents, when the science editor Susan Watts introduced her package by saying, 'No one disputes that Venter has made the crucial breakthroughs that mean he is now leagues ahead in the decoding game.' I was amazed, because this was so far from the truth and yet it was being accepted as a starting point for the discussion. As soon as the presenter, Jeremy Paxman, brought me on I said, 'The public program is actually ahead. We have already released two-thirds of the human genome.' It was now Paxman's turn to be amazed, and he said, 'If that's the case and it's all in the public domain anyway, what are we worrying about?' I answered that the Clinton–Blair statement was unlikely to make any immediate difference to the patenting of DNA because so much was already publicly available, but that it was a very valuable lead for future debates. I said that I thought most people agreed with the statement, and that it was a superb endorsement of the HGP effort—putting the genes in the public domain where they should be. Said Paxman: 'I couldn't agree with you more!'

A lot of people thought we came out of it rather well, but Susan Watts phoned up the next day and said it had been 'disappointing.' Jeremy Paxman is not supposed to agree with his interviewees; what the producers are after is passionate debate. I told her I was sorry, but she had altered the substance of the discussion by saying Celera was winning hands down. It wasn't especially her fault: she was simply repeating the picture that had been so carefully fostered. On the telephone with her beforehand I had been more passionate about the iniquity of patenting DNA sequences, but on camera I was forced to deal with her introduction. I really buttoned myself down, and exuded quiet confidence rather than the passion she was hoping for. It was altogether a fascinating insight into the workings of the media and the power of unchallenged press releases.

On 6 April, the day that Craig was once again to testify before a congressional subcommittee investigating the progress of the public and private efforts, Celera announced that it had completed the sequencing of the first human genome. It would assemble the sequence over the next few weeks, and was soon going on to sequence the laboratory mouse. Once again there was a most in-genious press release that talked about elevenfold coverage of the genome. That sounded very impressive, and appeared to be vastly more than the public domain had achieved. The correspondents, as intended, were immediately convinced that Celera had won. But we knew that such sequence coverage was impossible with the capacity that they had: we were all running neck and neck, and knew exactly what could be done in a given time. Rapid digging revealed that they were talking about clone coverage, not sequence. In other words, they had sequenced a few hundred bases at each end of enough clones to cover the genome eleven times, a step they used to build 'scaffolds' to help with assembly. Nevertheless many press reports gave credence to a confident Craig asserting that everything would be put together in six weeks, and as usual paid little attention to our comments on the changed definition of sequence!

A day or two later a HUGO meeting was held in Vancouver. Although it had in the end played little part in large-scale human genomics, HUGO had come into its own with its annual meetings, which are well attended and have become a major event in the calendar of professional activity. Francis was asked to speak on the HGP effort that year. During the questions at the end of his talk, a member of the audience asked why, if Celera had finished the genome, the public effort was continuing. Francis explained that the measure being used by Celera did not amount to completed sequence, and pointed out that the assembly would be difficult and would require finishing. 'You should not take at face value any claim by any group for at least two years that says we have finished the human genome sequence,' he said. 'It will not be true.' That was a truthful and sober statement, and I was delighted to see it in circulation. Francis was not denying what Celera had achieved, just explaining the reality of producing sequence.

Celera shares, which had shot up in response to its press release, plunged once again. Soon after Francis returned to Washington his department issued a partial retraction, declaring that he 'didn't say anything critical of Celera in his speech.' Following this episode, Francis pointedly refrained from making public statements of any kind about Celera.

I was appalled by what I interpreted as deliberate muzzling of the head of the Human Genome Project. It brought home to me forcefully that the strength of the industrial lobby in Washington means that no public servant can make statements that imply criticism of a commercial company (and of course, things are not greatly different in the U.K. and other industrialized countries). Francis agrees that the political reality of his situation was very different from my own.

The Sanger Centre has the support of Wellcome Trust, with whom they are philosophically very tightly aligned. That's a less complicated position than I find myself in here, where there is a great interest in seeing private-sector efforts in biotechnology flourish, and anything that appears to be in any way critical of that can potentially be problematic. So it has been possible for John to speak his mind at times very bluntly, when some of the rest of us had to watch our language extremely carefully in order not to set off alarm bells.

Celera's success in silencing Francis was very valuable to the company. To give one example, Columbia University in New York had been due to hear from him at a seminar in June, which he had to cancel (presumably he could have gone along and given an expurgated account that accepted Celera's claims at face value, but of course that would have been counterproductive for both science and the truth). Meanwhile, Gene Myers of Celera did give a seminar at the university, indicating that the human assembly was going immensely well, without giving much detail. The result was that most of the audience became convinced that Celera had done it. The same scenario, repeated across the country, established an aura of success that greatly helped the company in the build-up to selling its database.

Meanwhile Clinton was getting impatient with the continued brawling between Celera and the publicly funded scientists. The White House wanted something nice to happen about the human genome, which was now getting such a lot of press attention. It was a party political issue, with many of the Republicans supporting Celera, hoping to garner support from the biotech industry which had taken a beating in the markets, and many of the Democrats supporting the public side. Continuing conflict could be extremely bad for Vice-President Al Gore in his bid to succeed Clinton that presidential election year. Clinton sent a note to Neal Lane saying 'Fix it ... make these guys work together.' Ari Patrinos of the Department of Energy played the part of honest broker and got Craig and Francis round to his house for beer and pizza at the beginning of May. After a few more rounds of this 'pizza diplomacy', as *Time* magazine called it, they agreed to a joint announcement of the completion of the sequence, simultaneous publication later in the year and a truce in the war of words about who had done what. The negotiations went on in complete secrecy, to Francis's discomfort.

I felt pretty uneasy about doing that. It was clear that it had to be done under conditions of confidentiality or Craig wouldn't be willing to play ball. And yet as someone who is used to communicating with my colleagues at every little step, for a couple of weeks I wasn't able to do that, and that put me in a very awkward position.

All we knew was that Francis was lying low, refusing to do any press interviews. Everyone seemed convinced that Celera would announce the completion of the draft genome in June. Our official line was that although we also hoped to reach our target of 90 percent in June, we would keep any announcement low-key and go for a big song and dance around the time of publication, which we expected to be in September. It was only in early June that we found out what had been going on. In addition to the beer and pizza sessions between Francis and Craig, the British science minister, David Sainsbury, had visited the United States and talked to the White House science advisers about making a simultaneous United States–U.K., public–private announcement of the completion of the draft human genome. The date of the announcement, 26 June, was picked because it was a day that happened to be free in both Bill Clinton's and Tony Blair's diaries.

It was not clear that the Human Genome Project had quite got to its magic 90 percent mark by then, and Celera's data were invisible but known to be thin, so nobody was really ready to announce; but it became politically inescapable to do so. We just put together what we did have and wrapped it up in a nice way, and said it was done. We were sucked into doing exactly what Celera has always done, which is to talk up the result and watch the reports come out saying that it's all done. Yes, we were just a bunch of phonies! But we were trapped by Washington politics.

Later that day I went round to the Channel Four news studio to be interviewed by Jon Snow for the seven o'clock news. As usual I talked about free data release and its importance. Then they brought in Craig Venter over a transatlantic link from Washington, and asked him about patents. 'We never said we would patent thousands of genes,' he said, asserting that at most the number of patents they might be licensing to their pharmaceutical partners was in 'small double digit figures.' 'The fact that the bar has been raised [by a change in the United States patent office guidelines] is good news for everybody,' he added. It was tremendously different from the October 1999 press release with its 6,500 patent applications. It seemed to me that our actions in putting so much sequence into the public domain had indeed changed the business plan of the company, or at least the spin it wanted to put on its activities.

Of course, the 26 June announcement was a political gesture, but it genuinely didn't feel like that on the day. I remember thinking that it really had worked. It didn't matter that it was founded on the White House desire to get Al Gore elected. What mattered was that people were not talking about 'the race' anything like so much. They were actually talking about the implications of the work. And the Human Genome Project's sequence, incomplete as it was, really was available to anyone to use.

7 IN THE OPEN

IT WAS THE MOMENT OF TRUTH. WE HAD WAITED ALMOST THREE years to see if Celera would be able to make good its claim that it had sequenced the human genome faster, more cheaply and more completely than the public project. Now, in an article to appear in the journal *Science*, it was announcing that the goal was reached. 'A 2.91 billion base pair consensus sequence of the euchromatic portion of the human genome was generated by the whole-genome shotgun sequencing method,' the article began. With our own results due to be published the same week as Celera's, we were naturally eager to compare the two, and as previously agreed we exchanged papers with Celera shortly before the date of the joint public announcement, February 12, 2001. As we read for the first time through the detail of what the Celera team had done, it became clearer and clearer that the whole-genome shotgun strategy had not lived up to the claims made for it.

We couldn't quite believe it. We had fully expected their sequence to be better than ours, given that they had access to all our data and we knew that they were using it. But they were publishing a sequence that seemed overall no better than the publicly released sequence, and which depended heavily on it. Though this dependence was glossed over, it was there in black and white for anyone who chose to read the paper carefully enough.

Celera had been launched with the promise of quarterly releases of data culminating in publication in a leading academic journal; but as the Human Genome Project pursued its own draft strategy, the first undertaking was quietly dropped. As a result, the publication would be the first opportunity offered to anyone other than Celera's paying subscribers to evaluate what the company had actually achieved.

For a paper of such importance, researchers traditionally choose one of two journals: *Science*, the journal of the American Association for the Advancement of Science, and the British-owned journal *Nature* published by Macmillan. Both journals are international weeklies, but *Science* has a higher circulation within the United States and so is often preferred by United States scientists. Both had clear policies on publishing sequencing papers. As sequencing projects had grown larger and larger it had become quite impossible and pointless for journals to publish the sequences in print. So the authors of genome papers had to agree to deposit their data in one of the public databases (which in practice meant all of them, as GenBank, the European Molecular Biology Laboratory Data Library and the DNA Data Bank of Japan shared their holdings), where anyone could verify the findings outlined in the paper. Data in the public databases are free to anyone not only to read, but to download and analyze without restriction. A commercial company can even package and sell them if it so chooses. The databases act as universally available resources with the aim of advancing understanding and discovery for the benefit of all.

It was becoming quite obvious in the course of the HGP's abortive negotiations with Celera at the end of 1999 that the company had no intention of depositing its human sequence data in the public databases, as it had done (though somewhat under pressure) with the

Drosophila data. Instead, as it announced in a January press release, it would make these data available (exactly how was not clear at this stage, but from its own website seemed a likely possibility) with strings attached. One of the conditions was that users would have to agree not to redistribute the data. Commercial companies would also have to pay to make use of the Celera sequence.

Would *Science* or *Nature* accept a paper on conditions that fell so far short of their own guidelines? If either did, it would be a very serious matter for science as a whole. Scientific journals are critical to the integrity of the whole enterprise of science. They decide what does and what does not get published. Any article submitted for publication goes out to experts to be refereed, so that acceptance means that your peers have judged the work to be original and its conclusions valid. Of course, there are any number of examples of the peer review system failing—of brilliant work being rejected, or dubious results accepted—but, flawed though it is, the system more or less works. For the author, a published paper is a vital addition to his or her professional worth: quality and quantity of publications have become the main criteria by which a scientist is judged. Once a paper is out it becomes another brick in the hall of science, for others to build on or challenge. In some ways it's more like a termite mound, each of us industriously constructing little bits of the castle, adding extra chambers, repairing damage, knowing that later generations may build on our work or demolish it and build something different. But the enterprise succeeds only if our handiwork is in the open for all to see and make of it what they will. With so much at stake, journals have to act with the utmost probity in their dealings with scientific authors.

Part of the agreement with Celera, negotiated in Francis Collins's beer and pizza sessions with Craig Venter and Ari Patrinos, was that we would publish our papers simultaneously. The working assumption was that this also meant in the same journal, probably *Science*. Mark Patterson, one of the *Nature* editors, suggested to me that if the sequence paper did indeed go to *Science*, perhaps a complementary set of papers on mapping could go to *Nature*. Then both journals could share in a historic announcement. It was a good compromise that found general favor, and the two journals agreed to publish all the papers in the same week, whenever that should be, so that the whole effort would make a nice big splash. But we needed to know what *Science* (or indeed, *Nature* if *Science* rejected Celera's paper) was going to do about the data release issue. It seemed unlikely that *Nature* would bend its rules, but we weren't so sure about *Science*. We knew that the journal was negotiating with Celera. If it wasn't prepared to change its rules, there would be nothing to negotiate about. If it was going to make a special case of the Celera data, then I for one did not see how we could agree to publish in the same journal. This was an issue that went well beyond the antagonism between the public and private genome projects—it was fundamental to the practice of science.

Once they heard what was going on, leading figures from throughout the scientific community wrote to voice their concern. Aaron Klug of the Royal Society, Bruce Alberts of the National Academy of Sciences and Harold Varmus, ex-head of the National Institutes of Health, all joined the discussion. But the editors of *Science*—Floyd Bloom was succeeded by Don Kennedy during 2000—argued that when publishing private data they needed to recognize a company's right to protect its investment in sequencing. Kennedy responded to the concerns of the scientific community by claiming, not very effectively, that making the data available at Celera's website met the magazine's condition of deposition in a 'public database.' At the same time he came up with precedents for companies publishing in *Science* without revealing all their data— but none of them was a genomics company. The point about sequence data is that they are not just the raw material for the substance of the article, they *are* the substance of the article. And the argument for having all genomic data in one place is very specific. If you want to do any kind of serious analysis of the genome—finding genes, control regions or long-range features of any kind—you need to be able to access all the data at once. If Celera could argue for keeping its data separate, then others could do the same—in which case you would end up with a 'balkanization' of the genome sequence that would destroy its very purpose as a tool for discovery. Biologists would have to consult one source after another—and with the non-redistribution condition, they would not be able to incorporate the private data into the publicly available genome databases such as Ensembl that were being set up to provide easy access to the sequence and related data.

The political controversy surrounding Celera's submission was an unwelcome distraction from the much more important task for us of putting together our own paper on the draft sequence. The publication of the two papers would be a watershed in the history of the genome. The 26 June event had been a great day, but it would mean nothing to posterity—it left no trace behind but news reports relaying the hyperbole of politicians and scientists alike. It was dishonest to the extent that both sides had had to modify their definitions in order to say the working draft was complete. But the publications would be unambiguous. They would have to give the facts in full, making it clear just how far short of complete each sequence was. They would have been scrutinized by critical but informed colleagues, to ensure that vague points were clarified, weak claims strengthened or eliminated. Most importantly, they would pass into the scientific literature to be consulted again and again by scientists in future years and decades.

We began to think about our paper early in 2000, and Eric Lander drafted a first outline. It would include all the history and background to the project, so that anyone outside the field would really understand how it was done. But the publication would make no difference to what people did every day in the twenty sequencing centers worldwide whose work was being documented—they would keep adding, every hour, minute and second, to the total number of bases sequenced, and every twenty-four hours that new information would arrive in the databases. So we would have to choose some essentially arbitrary cut-off point at which we would freeze the data, to provide a snapshot for our analysis. We did this for the 26 June announcement, but for the publication we would be able to choose a later date and so have a more complete set of data. This was an extreme case, but it's actually true of a lot of science that the exact publication point is arbitrary. Most research is work in progress, with new questions arising from every answer and the concept of a breakthrough more often than not an artificial one. But having a clear written record in terms of milestones is nevertheless tremendously valuable.

The sequence was accumulating in the databases. Assembling the sequence was a far from straightforward task—because many of the data were based on unfinished clones with gaps and errors in them, because sequence was being collected from more than twenty different sources and because the high proportion of repeat sequence in the genome was a continuing headache. Jim Kent at the University of California at Santa Cruz, who was then still a graduate student (but one with a previous life in the computer animation industry), spent a month writing a huge piece of code called GigAssembler to do the job, which ran successfully for the first time four days before the 26 June announcement of the draft. Ewan Birney, Tim Hubbard and their colleagues at Hinxton were steadily improving the capacity of Ensembl to predict genes using the full range of bioinformatics tools. Along with the map at St. Louis, all these sources of genome information were freely available online, and would be described in the paper. But as time went on, ambitions for the scope of the paper grew, with some wanting it to include much more detailed analysis of the genome. On the whole, I was for an earlier and briefer account, but I could see there was a case for more comprehensive treatment. I was vaguely aware that Francis

Collins was continuing to negotiate with Craig about how the publications were handled, and one element of this discussion seemed to involve a later publication date that would give us more time.

Eric Lander started to put together an international team of bioinformatics people, to form what he called the 'hardcore analysis group.' Coordinated through regular meetings, conference calls and a blizzard of e-mails, this group steadily produced the figures, tables and diagrams that would reveal what the sequence was really telling us.

The paper was a key topic for discussion at the eighth international strategy meeting, held at Evry near Paris in September 2000. It would reflect the international nature of the Human Genome Project in both authorship and content. There would be no names on the title page—the author would be simply the International Human Genome Sequencing Consortium. Twenty centers would be listed as members of the consortium: twelve from the United States, five from Europe (one each from the U.K. and France, and three from Germany), two from Japan, and one from China. Other centers that had sequenced small regions were listed in the acknowledgements. We also wanted to discuss with our international colleagues the still unresolved position with respect to *Science* and data release. There was general agreement that if *Science* compromised unduly on data release, we would move the paper to *Nature*. After the main meeting, Francis Collins convened Eric Lander, Bob Waterston and me to discuss progress. We agreed that the four of us would take responsibility for seeing the paper through to completion. Eric, who writes well and enjoys doing so, would take on the final editing and harmonization.

A month later I joined the hardcore analysis group in Philadelphia as they finalized their results. It was indeed exciting to see the analysis emerging, the 'landscape of the genome', in the phrase that was on everyone's lips. On the Saturday afternoon that same weekend, Eric and Francis went off to the annual meeting of the American Society of Human Genetics to make their speeches and accept the Human Genome Project's share of an award for sequencing the human genome. The other share was to be accepted by Craig Venter. When they came back they were subdued, as well they might be. It was so extraordinary. Everyone could see the public data; nobody could see Celera's. Yet an award was being given on the strength of the company's statements. Had this ever happened before, we asked ourselves at the Sanger Centre when we heard of it? That an internationally reputable society would give an award for research that was unpublished and unseen? This had nothing to do with what one believed to be the actual facts of the case; that was a separate issue. The shocking thing was that at this stage there was no evidence on the basis of which they could make the award at all. First science by press release, and now awards by press release!

Two days earlier, Jane Rogers had seen Francis at the conference, and had asked him if he didn't feel that the situation was unethical. 'Jane,' he had replied, 'there are no ethics in this.'

The four organizers of our paper—Francis Collins, Eric Lander, Bob Waterston, and I—were united in our opposition to *Science*'s position on the release of the Celera data. Expressing the unanimous view of the Sanger Centre, I was for moving to *Nature* immediately; if the Celera data were not going into the public databases that was all we needed to know. But those at the American end were more cautious. They were equally unhappy about any retreat from the usual data release conditions, but once again, political conditions made it difficult to take a decision that clearly implied criticism of United States industry. They felt they could not act until we had seen the wording of the conditions that were to be placed on access to Celera's data, and continued to hope there would be time to negotiate. 'The devil is in the detail,' they said. On the contrary, thought I, the devil is in the principle.

The months went by—the goal of September publication had long since gone by the board—and at the end of November we still did not know exactly what sort of deal Celera had struck with *Science*. Michael Ashburner, the Cambridge *Drosophila* geneticist, joint head of the European Bioinformatics Institute and a staunch supporter of unrestricted data release, wrote an outraged letter to every member of the magazine's board of reviewing editors, of which he himself had been a member until not long before. He told them he was refusing to review any more articles for *Science*, or to submit any to the magazine if *Science* went ahead with the Celera paper on the existing basis, and he urged his former colleagues to follow his example and resign. Copies of the letter circulated rapidly on e-mail, and the issue became a topic of heated debate in lab coffee rooms, although surprisingly it barely surfaced in the press.

Eventually it was all over very quickly. We finally got to see the text of the material transfer agreement that commercial companies would have to sign if they wanted to look at Celera's data, and the restrictions on academic users. The latter would be able to download 1 megabase per week by clicking on the Celera website, subject to a non-redistribution clause; if they wanted more they would have to get a signature from a senior member of their institution guaranteeing that the data would not be redistributed. The restriction on redistribution meant in effect that no public sequencing lab could even look at the data without laying itself open to a potential lawsuit. Commercial companies would have to pay to use the data, and were also bound by an agreement not to redistribute them; companies Eric consulted suggested that many of them would not be able to sign. There was no time left to discuss any modifications to these agreements, which we found completely unacceptable. After a rapid circular to all the members of the sequencing consortium to confirm their approval, Eric wrote on behalf of all of us to the editors at *Science* and told them that we were submitting our paper to *Nature* instead. It finally went off on 7 December.

The same day Daphne and I flew to New York to visit our daughter Ingrid and her husband Paul Pavlidis. During her Ph.D. in Berkeley, Ingrid had worked as a volunteer at the Exploratorium in San Francisco and discovered a passion for science communication. She and Paul had married and moved to New York, where she now had a job developing exhibits in bioscience for the New York Hall of Science, while Paul was working in bioinformatics at Columbia University. They had a university apartment nearby. As evening fell on that day, Daphne, Ingrid and I left their apartment to go shopping, turning away from the cathedral and down the street to the west, crossing Broadway with its shops and bustle and on down to the quiet boulevard above Riverside Park. We walked south, the lights of New Jersey shining across the dark waters of the Hudson River. Above them the blue of the sky deepened, with pink streaks of cloud down river. Drifts of dry leaves crunched under our feet, the row of empty benches that had lost their occupants for the winter snaked on under the great trees.

Ingrid was heavily and beautifully pregnant with her first child—our first grandchild. Walking with the two women, I thought of the genetic events that bound us together so closely, of the three generations of human beings through whom one strand of the common thread of humanity was being transmitted, and of the events, both stirring and sad, of the years of endeavor to decipher the code.

On our return to the apartment I opened my laptop and logged on to my e-mail. The download of the last message, over the phone line from Hinxton, took for ever. But at last there it was, from Mark Guyer at the National Human Genome Research Institute to Carina Dennis at *Nature*, with the huge attachment and the header: 'Carina: On behalf of the International Human Genome Sequencing Consortium, we are very pleased to be able to submit the attached manuscript for publication in *Nature*.' We had taken a stand for freedom of information and integrity of scientific publication. The scientific world would be made aware that what *Science* had done was unacceptable to us, to our advisers and (we guessed) to most of our fellow scientists. We would encounter criticism from some who felt we had overreacted, but we could face them with confidence knowing that in the eyes of the majority we had done the right thing.

Later, I walked out in the freezing air again, helping Paul to collect a new sofa. I was chattering away about the submission and how important it was. Then I checked myself. Why? Why was I justifying the action to Paul? What I found, with him as with everyone who was not directly involved, was that nobody knew what was going on—or didn't believe it. And I reflected, not for the first time, on the power of public relations. Like those who can afford expensive lawyers, those who can afford expensive PR usually get their way—or at least, exert influence beyond what is justified. The penetrative, unremitting power of Celera's PR had so convinced the newspapers, and through them everyone, including many of my fellow scientists, that my own truths counted for nothing. Once a particular point of view has taken hold in the public imagination, it's extremely hard to offset it. The only recourse is to compete on the PR front in the first place. I find that a profoundly depressing thought. Is it a fantasy that simply being honest will in the end be powerful enough?

In the middle of the frantic effort to get the paper together, I finally stepped down from the directorship of the Sanger Centre. As I had expected, the Wellcome Trust had been in no hurry to act on the notice of resignation I had delivered almost two years before. But eventually they accepted that, with the sequencing of the human genome well on target, it was not a crazy idea to find a director with different aims and skills who would extend the Sanger Centre into the era of functional genomics—using the genome to understand biology. They found Allan Bradley in Texas. Allan had begun his career in genetics in Cambridge, but had moved to Baylor College of

Medicine in Houston in 1987. There he had carried out pioneering studies of development in the mouse, using gene knockout techniques to explore how the process is controlled. Still only in his early forties, Allan was well placed to encourage new perspectives while sequencing continued on other species including the mouse, the zebrafish and many pathogens.

Allan was to take over at the beginning of October 2000. On the Saturday following the Paris meeting Bob Waterston, Rick Wilson, John McPherson and I travelled back to London by Eurostar, and Bob and I went on to Stapleford and a warm welcome from Daphne. Bob seemed a little preoccupied, and kept scribbling on a big yellow pad. On Monday we went into the lab, and again every-thing was a bit strange. I knew there was to be some kind of farewell do, but no one had told me the details and I thought maybe there would be a few drinks at the end of the day. I tried to set up meet-ings for the afternoon, but kept being met with shifty evasions. At the end of the morning I was seized and taken to the Garden Room (now the James Watson Pavilion) where all the senior staff had gathered for a splendid lunch.

From there we went to the auditorium and I was immensely moved as one by one my closest colleagues came to the lectern. First—and now I knew what the yellow pad was about—was Bob. To my amazement, the first story he told was the story of Syosset: how we had stood on the station platform and realized the enormity of what we had undertaken, and how I had said I heard the prison door closing. We had never discussed that conversation since, and yet both of us had retained the memory as a pivotal moment in our relationship and the story of the genome. A parade of other key figures followed. After more drinks, I was marched back to the auditorium and discovered that the metaphorical curtain was about to go up on a full-scale pantomime. Christmas pantomimes had been a regular event since the Sanger Centre started; John Collins and Ian Dunham were the principal scriptwriters. This time they had surpassed themselves. 'King John and the Knights of the Holy Genome' featured all the main characters in the human sequencing story, but owed more to Monty Python than Thomas Malory. Two minutes or so into the show I was hauled on to the stage to play the lead, with Jane Rogers handing me flash cards for my lines. With interludes from an ABBA lookalike singing group ('He is the sequence king...') and a Big Brother-style video ('I think John should go because...'), it was the most amazing honor: a personal Sanger Centre pantomime.

I'd had plenty of moments when I wished I wasn't director, but never because of my colleagues. The ethos among them is fantastic. There's a huge sense of team spirit. At the same time everyone is serious about what matters, and works with single-minded commitment. If anyone had ever had any doubts about that, they were instantly dispelled at the end of October, when we got our share of the autumn floods that submerged much of Britain in 2000. Our architect-designed lab is built on the flood plain of the Cam. 'Don't worry, it'll only flood every hundred years,' the architect had said. Four years after the building was completed we had two feet of water in the basement of the West Pavilion, which we had filled with sequencing machines in order to scale up our human genome effort. Almost as one, the sequencing teams swung into action and moved all the machines to a higher floor. Allan, who had been in post for less than a month and had just left on a visit to Houston, returned to find that within two days sequence was flowing once more. There was no panic, no argument, just teamwork and dedication.

Was I sorry to be leaving that behind? Well, for a start I wasn't leaving, just moving to a smaller office—until the paper came out I would still be busy with the Human Genome Project, and I had unfinished business with the worm. I'd still be able to enjoy chats with everyone. But I'd always been a reluctant director; I'd come close to resigning before over issues that I felt to be important. So not being director any more was a relief more than anything. People kept asking what I was going to do next. But I seemed to be as busy as ever.

After our decision to move the paper to *Nature*, early February 2001 became the target date for publication. Despite our decision not to submit to *Science*, the two journals upheld the agreement to publish the Celera and HGP papers simultaneously. When we heard that the publication date had been moved back by another three weeks, we were furious. We were ready for early February, and so was *Nature*. The delay gave us no advantage at all—but it could provide a huge advantage to Celera over and above getting its paper ready. The American Association for the Advancement of Science, which owned and published *Science*, was holding its huge mediafest of an annual meeting from 15 to 20 February. There was to be a special weekend symposium on the genome organized by Craig in association with *Science*; Francis would also be giving a keynote lecture. Under the terms of the usual embargo agreement *Nature* imposed on its authors and the scientific press, we would not be able to give a preview of our findings. But what Celera might come up with was anyone's guess. I could not see it as anything other than a spoiling exercise. Fortunately we were able to prevail on *Nature* and *Science* to settle on the week before the AAAS meeting for publication, with a joint press announcement on 12 February and an embargo on press publicity until then.

It was during the build-up to the press announcement that we finally got a chance to look at the Celera paper. What was immediately obvious was that Celera had made no attempt at all—or rather, no attempt that they were prepared to show publicly—to assemble their own sequence data alone. So what was Celera publishing, if not their own data and assembly? As they had announced early in 2000, they had abandoned shotgun sequencing when they reached just over fivefold coverage—about half their original target—and instead incorporated the public data into their data set. For the paper, they had then produced two alternative assemblies of the data. One they described as a whole-genome shotgun assembly.

This included an extra 2.9-fold coverage from the publicly available Human Genome Project consensus sequence, the data shredded in the computer into pieces the size of normal sequence reads, which the Celera authors referred to as 'faux reads.' But these reads were not selected at random—they formed a fully overlapping set. So even though they were chopped up and scrambled before being fed to the Celera assembler, they still contained all the information necessary to put themselves back together in the right order, and recover the sequence generated by the HGP. Furthermore, the 'whole genome assembly' also involved a process described as 'external gap walking,' which meant filling gaps with assembled BAC data from the HGP. The paper itself revealed that this is what was done, but only if you knew where to look—the whole impression you got from the introduction was that the assembly was a triumphant vindication of the whole-genome shotgun approach.

The second version of Celera's assembly, which they called a 'compartmentalized' assembly, made explicit use of the HGP map to put the sequence in order. There was no ambiguity this time—the compartmentalized assembly depended on the work of the public project, and the authors acknowledged this. What was interesting was that they went on to use the compartmentalized assembly, and not the 'whole-genome shotgun' assembly, as the basis for the analysis of the genome that took up the rest of the paper.

In addition, because of the public availability of our draft, they were able to publish a comparison of the two products as part of their paper. Not surprisingly, this comparison was not presented as being in our favor. We would not get an opportunity to make our own public comparisons until the day of the announcement on 12 February.

Eric Lander, Richard Durbin and Phil Green all independently analyzed the information and came to similar conclusions. There was no evidence in the paper that the whole-genome assembly had worked adequately, and the compartmentalized assembly seemed overall comparable with what our own effort had achieved. Phil's view was particularly telling: he had kept out of the consortium in order to maintain his independence, although of course we all used his assembler, phrap, to put our clones together. Phil had refereed our paper for *Nature*. Referees are conventionally anonymous, but Phil made a point of signing his referee's report. His initial review of the international consortium's paper consisted of fourteen closely typed pages of comments. We'd been kept busy all over Christmas dealing with his points. None of it was destructive; he'd just pointed out where we'd been sloppy in reporting and needed to do some tidying up.

Despite disparaging remarks from Craig Venter about 'pissing contests', our intensive analysis wasn't carried out in the interests of my-contig-is-bigger-than-your-contig point-scoring. Rather, it was the first step in discussing objectively, rather than by press release, what had been achieved by the various approaches. From the evidence of the paper it appeared that without the public project there would not only have been no publicly available draft human genome by 2000—there would have been no draft genome at all. More seriously, the chances of ever having a fully finished sequence would have been very slim indeed.

There was no chance that Francis Collins would be able to point out these uncomfortable truths at the joint press conference with Celera held in Washington on 12 February. But Eric Lander had already given an on-the-record, embargoed briefing to American journalists almost as soon as he saw the Celera paper. And I had every intention of telling the British press exactly what had and had not been achieved at the London press conference *Nature* was organizing at the Wellcome Trust headquarters. Just to be on the safe side—we thought it highly likely that the Celera publicity machine would come up with some kind of diversionary tactic—

Richard Durbin and I also briefed the senior London science correspondents the previous Wednesday. I gave an introduction, then Richard took them through slides showing the assembly methods and showed a comparison table. He presented our conclusions: that the whole-genome shotgun had not worked as claimed; that Celera had used more of our data than it admitted; and that it had come up with a product that was in some respects better than ours and in some respects worse. Our additional objective was to make sure that nobody thought the sequence was finished, even though the Celera authors studiously avoided calling their assembly a draft.

As soon as it was over the journalists all headed off to Lyon to attend Biovision, a big public conference on biotechnology. Craig Venter was speaking, and it transpired that they all had interviews with him set up for the Friday afternoon. On Saturday morning Robin McKie rang me at home. Robin is the scientific correspondent of the Sunday *Observer*, and had not been invited to our Wednesday briefing or given any of the advance information distributed to the press by *Nature*, which was embargoed for Monday. *Nature* tends not to include Sunday newspapers in its distribution of embargoed advance information, because they are notorious for disregarding embargoes—Robin himself had jumped the gun with the story of Dolly the cloned sheep four years previously. He argued that he had researched the story from other sources and so was not bound by *Nature*'s embargo.

On this occasion, fortunately, Daphne picked up the phone. I was still in bed, hoping to get rid of a streaming cold that had plagued me all day on Friday. She had a chat with Robin and came back saying, 'He's talked to Craig, and says it's very important that the public side answers.' Daphne had handled the situation adroitly, and had gleaned from Robin that Craig had made a statement. I didn't phone Robin back—that would have been a breach of the embargo on my part—but phoned up the Wellcome Trust's press office team and told them something was up.

Late that evening we learned that Robin's interview with Craig was to be the front-page story of the next day's *Observer*, and was already out on the web. *Science* magazine declared that the embargo had been breached and released its paper. *Nature* had no choice but to do likewise. The biology editor, Richard Gallagher, was very fed up. We all felt that the story did not amount to a breach of the embargo. The only substantial information from the papers that it contained was the estimate of the total number of human genes. Both groups had put the number in the range 26,000–40,000, less than half the figure of 100,000 that had been bandied around until very recently. It wasn't even a very new point—estimates of gene numbers had been coming down steadily, and certainly other papers, such as Ian Dunham's on chromosome 22, had been published giving this kind of figure a year earlier.

What Craig did in the *Observer* interview, however, was to use the number to make a bogus philosophical point: that the small number of genes implied a much greater role for environment in determining our natures. No longer, he said, would it be realistic to assume that there were specific genes for behavioral traits such as thrill seeking, intelligence or athletic ability. 'We simply do not have enough genes for this idea of biological determinism to be right,' Craig was quoted as saying. McKie called the finding 'a radical breakthrough in our understanding of human behavior.' This hyperbole was entirely unjustified. The fact that we have only twice as many genes as a worm or a fruit fly is extremely interesting biologically, but it adds nothing at all to the nature–nurture debate. We already knew that genes, especially control genes, combine their actions in ways that we have barely begun to understand—a theme to which I will come back later.

Of course, there were recriminations about the *Observer*'s action. Robin said that he interviewed Craig Venter privately in Lyon and that Craig understood that the *Observer* was not bound by the embargo. Celera denied this. But given that Celera has always given the impression that its management of public relations is second to none, it seems highly unlikely that Craig did not know what he was doing when he talked to Robin McKie. One of the journalists at Eric's press briefing had taped it and given the tape to Craig, so Celera knew that we were going to criticize the whole-genome shotgun results. Intentionally or not, the *Observer* certainly provided a diversion. I had to spend most of that Sunday giving interviews, and a recurring line of questioning was on the nature–nurture issue. (A day or two later I found myself answering the same question live on a radio program broadcast in Bogota.)

And so to the Monday press conference. To our relief the lecture theatre at the Wellcome Trust's headquarters was full, even though many of the papers had run stories already. We really had just one story that we wanted to ram home: that, thanks to the publicly funded Human Genome Project, the human genome was available to all, including scientists in developing countries. In his opening statement, Mike Dexter pointed out that scientists from the developing world had accessed the databases more than 300,000 times in the previous few months. I explained how Celera had made use of our data, but only so that I could point out that we simply would not have the sequence had it not been for the public project. Meanwhile, in Washington Bob Waterston and Eric Lander were doing their best to make the same points while sharing a platform with Craig. Eric thanked Bob for creating the genome map, saying pointedly, 'I am sure I speak for all of us when I say how grateful we are.' And Bob made the data release point: 'Most importantly,' he said, 'we have made this available to the world without any constraints. No patents filed on the raw sequences. No licenses. No documents to be signed. All you need is an internet connection.'

I'm not sure how much of it really sank in. Most of the coverage presented our arguments as just an extension of the bad-mouthing that had gone on the previous year, and Craig was eager to agree that it was all just sour grapes because he had 'stolen our thunder.'

With a few honorable exceptions—Aaron Zitner of the *Los Angeles Times* was one, and the *New Scientist* another—hardly anyone acknowledged that the principle of free release was a moral imperative on our side. If free release was mentioned at all, it was presented as one of two equally valid alternative choices.

The struggle to keep genomic data public and free continued throughout 2001. A consortium consisting of three private companies, six institutes of the National Institutes of Health, and the Wellcome Trust had been formed in October 2000 to produce a draft sequence of the genome of the laboratory mouse. The mouse genome is important both in its own right and because it helps enormously with the interpretation of the human genome: wherever you have a match between mouse and human sequence, you are likely to learn something about how nature makes a mammal. By May 2001 the Mouse Sequencing Consortium, which was releasing its data freely (as we did for the human sequence) on its course towards a finished mouse sequence, had a draft with three-fold coverage.

Having taken advantage, as it was entitled to do, of the publicly released human sequence, Celera was able to switch to sequencing the mouse at an even earlier stage. In July 2001 it released a press statement saying that it had assembled a draft with five-fold coverage, two-fold more than the public–private consortium. Consequently some labs paid for access to the Celera mouse database, though at the same time they declared that they were eager for the freely available finished sequence that would eventually be produced by the consortium. As long as there are both freely released and proprietary sequencing efforts running in parallel there will inevitably be more data in a proprietary database (since it contains data from both) until each genome in turn is finished. Again, as in the case of the human, there was a danger that this position might offer a route towards monopoly: Craig Venter called (unsuccessfully) for public funding for mouse sequencing to be terminated just as he did previously for the human.

Other than the press release there was no publication on Celera's mouse genome, so nobody could interpret how the company had assembled its data. Meanwhile, Bob Waterston, Eric Lander, and I wrote a brief analysis of the extent to which Celera had used the HGP data to generate both assemblies of the human genome reported in its Science paper. Aaron Klug, as an independent observer, communicated our paper to the *Proceedings of the National Academy of Sciences,* and it was published in March 2002.

The following issue carried a vigorous rebuttal from Gene Myers, Craig Venter, and their colleagues, in which they attempted to show that HGP data played little part in their assembly process. They didn't dispute most of the analysis in our paper, but focussed narrowly on the initial step of the process. It is true that due to the peculiarities of the Celera assembler, some of the assembly information inherent in the 'faux reads' taken from the public data would have been lost at this initial stage. However, enough would remain to provide useful short-range continuity, and no one appears to dispute that the later steps, 'external gap walking' and anchoring assembled fragments to the genome, necessarily use information from the HGP. Alongside the Celera reply, PNAS ran a second commentary on our paper from Phil Green. Phil not only agreed with our analysis, but went much further in questioning the claims of speed and efficiency made for the whole genome shotgun approach.

This is really the issue. Nobody is questioning the fact that whole genome shotgun can get you a lot of data, and it has always been used for simple genomes. But if you want to end up with a fully finished sequence you need some way to close all the gaps and resolve ambiguities. For large, complex genomes such as the human a clone map is essential given the current state of the art. The Mouse Sequencing Consortium adopted a hybrid strategy, combining whole genome shotgun assembly with clone-based sequencing.

Many commentators, cued by Celera, said that the race accelerated the timetable of the Human Genome Project—some said by as much as ten years, which is certainly incorrect. My own conclusion is that it will have made little or no difference to the date of the finished product (2003), though it did result in an intermediate draft being formally announced in 2000, in addition to the informal release of unfinished sequence that was happening anyway.

My reason for saying this is quite specific. In 1996 the Sanger Centre was funded to sequence one-sixth of the genome to fully finished standard by 2002. In 1998 we were on course, and by May 2001 we had already finished a sixth, in spite of the disruption caused by the draft activity and the associated PR distractions. When we were discussing this the other day I said to Eric Lander, 'I hardly think that you would have sat on your hands and watched us do it!' And that, of course, is the point: the internal competition that I've described within the Human Genome Project is not as destructive as it may sound. On the contrary, it ensured that we were all racing along, just as Bob Waterston and I had raced on the worm. So if we at the Sanger Centre were doing one-sixth by 2001 or 2002, you can be sure that Bob would try to do the same and so would Eric. So the pressure would have been on, a bit more funding would have come through, and we would be in the same position as we are today—or better, because of the lack of distractions.

But who cares? It doesn't do to get too upset about all this. In the fullness of time, maybe it won't matter that we did poorly on the PR front. What matters is that, as Eric puts it, 'the good guys won'—we produced a sequence, put it in the public domain and made it impossible for any individual or company to control access to it. And we will go on and finish it to the high standard we set ourselves at the beginning.

Maynard Olson and Phil Green, who opposed the draft strategy in 1998, still worry about this. In an accompanying article in the same issue of *Nature* as the sequence paper, Maynard said he feared the publication of the draft would lessen the motivation of people to finish the job. 'Each new round of press conferences announcing that the human genome has been sequenced saps the morale of those who must come into work each day and do what they read in the newspapers has already been done,' he wrote. The truth is that both the finishers at the Sanger Centre and their funders the Wellcome Trust are totally committed. They don't need to be told that they've got to do it, and they don't need to be told that there's a danger they might not do it. It is happening. That's one of the reasons why I felt OK about stepping down as director. If the funding had been in any way insecure I would have hung on. It may be that some of the other sequencing labs will move on to other species rather than putting the effort into finishing the human, but the remainder will have no trouble finishing everything by 2003. Indeed, as I write nine-tenths of the genome is up to finished standard. And I am absolutely with Maynard Olson in believing that finishing is essential. Errors need to be cleaned up and gaps closed if we are to compile an accurate catalogue of genes and unravel the mysteries that lurk in all that so-called junk.

By 2003 a fully finished sequence will become universally available, a work of reference as indispensable to biologists as a dictionary is to a writer. In the meantime, what have we learned from the draft?

Let's first go back to the issue of the number of genes. We think that the human genome codes for some 30,000–40,000 genes, only about twice as many as it takes to make the nematode worm or the fruit fly. But the claim that this should change our thinking, because 30,000 is too few to code for all our characteristics, is based on the assumption that each characteristic, from hair color to your level of interest in football, depends exclusively on one corresponding gene. The implication of the *Observer* article seemed to be that higher estimates, of, say, 100,000, *would* be sufficient to account for humanity in this way, so that reducing the estimate to a third meant that we should attribute greater importance to nurture in our development.

The notion of one character: one gene is largely false, but leaving that on one side for a moment, let's first note that the effect of both genes and nurture is already self-evident. Most powerfully, studies of identical twins show remarkable correlations in many characteristics, both physical attributes and behavioral tendencies, even when the twins are brought up separately. But the twins are not the same individual. Not all steps in development are tightly controlled, so that there are opportunities for random variation and effects of the environment. The result is that the twins end up with minor physical differences, but more importantly they have their own experiences, thoughts and ideas. They are unique human beings, even though they share the same genome.

The question is: Should it make a difference to our thinking whether we have 30,000 or 100,000 genes? I think the answer is no.

First, we simply know too little about how genes actually work. Having the complete gene set, which we are approaching through the genome sequence, will be of great help, but each gene has now to be painstakingly examined to identify its role. The gene list will be constantly scrutinized by people who are looking at systems in the body. The list provides them with help in finding all the components of the system—and that's all. In the long run this approach, taking apart the machine mechanism by mechanism, will narrow our area of ignorance so that we really can evaluate directly the relative roles of heredity and environment in a new way. The more precisely we understand how the machine works intrinsically, the better we can deduce the contribution of extrinsic factors. But we have a long way to go.

Nevertheless, at first sight the conclusion that it takes only twice as many genes to make a human as it does to make a tiny worm or a fly seems to lead very naturally to the conclusion that the genes are fairly boring and don't really have a lot to do with the essence of being a human. After all, runs the argument, a human is obviously so much more complex than a fly that twice as many genes just won't be enough. When I hear this argument I tend to hear also a subtext to the effect that humans, and the speaker in particular, are so much more important than flies that twice as many genes won't do.

What tends to be forgotten is management. Many of the extra genes that are added in going from worm or fly to human appear to be control genes, and they come in hierarchies which are only just beginning to be worked out. So in principle, by elaborating the control mechanisms, a huge range of tissue types can be specified and a very complex structure can be built up. It's a bit like the expansion of an organization: although some of us wish it wasn't so, an essential part of building up a large organization is the introduction of more complex management structures and the employment of more executives. The control genes are the executives of biological development, and they allow complex and diverse structures to be built from units that are fundamentally quite similar. Many control genes operate by switching groups of other genes on or off; in addition, a single gene typically gives rise to a variety of products. One gene can have two or more differently spliced RNA transcripts, acting as templates for different proteins, and enzymes can further modify the protein after it has been synthesized; all of these processes offer further opportunities for control.

But the most remarkable thing is the power of genes in combination. Consider a gene that can exist in two variants, or alleles: A and B. This single gene will then allow us to specify two cell types, A and B. Now add a second gene, that can exist as C or D. Together the two genes will allow us to specify four cell types: AC, AD, BC and BD. Three genes can specify eight types, four, sixteen types, five, thirty-two types; ten genes, over 1,000 types, twenty genes, over a million. The conclusion is that just a few dozen genes, if applied in a hierarchical executive fashion, can provide an immense amount of additional complexity. So the addition of an extra 15,000 over the worm's allowance allows plenty of room for maneuver in the construction of a human being. In real life things will not be as neat as this simple discussion suggests, but by thinking in this way we can avoid being trapped by falsely limiting assumptions.

By the same line of argument, we can tackle the concern that an extra 15,000 genes are too few to explain the range of human inheritance. Indeed they would be too few if each gene were solely responsible for the specification of one recognizable human characteristic. But we have long known that this is the exception rather than the rule. In particular, many of the subtle human attributes about which we care most—intelligence, athleticism, beauty, wisdom, musicality and so on—are clearly not heritable in the same way as hair or eye color, for example, leading some to the conclusion that they are not heritable at all.

But think again about the power of different alleles combining in different ways. Extending our range to thirty-three genes allows us to generate over 8 billion different types, enough to give every living person a unique label. Three hundred genes will provide as many different types as there are particles in the universe or seconds since time began—vastly more than will ever be needed uniquely to identify every person that has ever lived or will live. And this is on the parsimonious assumption that each gene comes in only two forms, whereas in fact there are numerous alleles of every gene. No wonder identical twins are so special: we shall never see two identical human genomes by chance, but only by the splitting of the fertilized ovum.

Taken together, combinations and hierarchical control allow us to see in principle how both the complexity and the diversity of humans can be specified by a relatively small number of genes. All this should give us optimism in moving forward to find out exactly how it all works; but it should also caution us against jumping to quick conclusions. The complexity of control, overlaid by the unique experience of each individual, means that we must continue to treat every human as unique and special, and not imagine that we can predict the course of a human life other than in broad statistical terms.

The genes are the starting point for a human being, and we should think of them as offering potentials rather than exercising constraints. Many fear that genetic information about individuals will be used to discriminate against them, and this is a concern that has to be taken seriously. Insurers are pressing to be allowed to use the results of genetic tests taken by their clients in deciding whether or not to issue policies; in future both insurers and employers, if the law permitted, might make genetic testing a condition of issuing a policy or offering a job. It is immensely important that we do not make presumptions about a person's health or ability on the basis of their genotype, but rather look to see what they can actually achieve. It is a matter of fundamental human rights: rights that are broadly accepted, at least in principle, in Western society as far as sex discrimination and race discrimination are concerned. The same rights must now encompass all forms of genetic discrimination, because we are acquiring the ability to measure a vastly greater range of characteristics than before. Although the correlation of genetic characteristics with physical and mental outcomes will in most cases be purely statistical, there will be a temptation—there already is a temptation—for these to be used in actuarial prediction, very possibly to the detriment of some individuals' opportunities. This we must oppose.

The new genetic knowledge is an enormously valuable starting point for research in biology and medicine. That is why it is so important to finish the sequence, so that it is as useful as possible. It is a permanent archive to which scientists will keep referring. But we need to be cautious about the immediate claims we make for it. Headlines such as 'Gene code could beat all disease' lead only to disillusion when year after year people continue to suffer from cancer, heart disease or senile dementia. To a certain extent I condone the hype—but only because it's important to keep the topic in the public eye so that there is wide debate about issues such as genetic discrimination. But let's think for a moment about what is really likely to emerge in the next few years, moving up the scale from easiest to most difficult.

The most immediate application, already well under way, is in diagnosis. Once a variant gene has been found that is associated with a particular disease, it is a trivial matter to conduct a test that tells you whether or not someone has that variant form. Genetic tests are now available for a number of diseases including cystic fibrosis, muscular dystrophy, certain forms of breast cancer and Huntington's disease. These are mostly comparatively rare conditions in which a single genetic defect gives you a high chance of developing the disease. A positive test result leads to hard choices for affected people: if it is a prenatal test, then parents need to choose whether or not to terminate the pregnancy; a woman with a positive genetic diagnosis of a predisposition to breast cancer, even if she has no tumors, may opt for preemptive surgery. Such choices are never straightforward, and patients need careful counselling to help them decide.

With the advent of the SNPs database, we can begin to get a handle on common genetic variations that have a more statistical impact on common disorders such as heart disease, asthma and diabetes. This is going to be more complicated, because no single variant will by itself make much difference to your susceptibility to disease—it will be a matter of groups of variants working in concert, and this is a very active research field at the moment. There is still a lot of work to do to correlate SNPs with illness in large populations, but undoubtedly genetic tests will emerge from this work and will be patented. When people patent a gene, all they mean is that they know its sequence and that they can use it to do a diagnosis. To me this should not be a basis for patenting the gene as a whole. I think we get into all kinds of trouble by laying claim to whole genes in order to protect rights to a diagnostic test, when what we really need are treatments for diseases based on those genes, which will take longer to develop.

The same knowledge of variation can contribute to improved drug treatments. It's a constant source of frustration for doctors that drugs that work very well on one patient, such as steroids in asthma, don't work at all on another. A SNP profile might be able to provide guidance to doctors on which is the best drug to prescribe. Further down the line, drug companies will undoubtedly start producing families of drugs 'personalized' for different SNP profiles. Whether the benefits of doing this will be worth the considerable costs remains to be seen. We have yet to discover whether the savings made by not having to try out a range of drugs until you find one that works will outweigh the expense of carrying out genetic tests on all patients.

The genome will also undoubtedly have an impact on people's choice of diet and lifestyle. In consumerist Western societies this will no doubt be seen as a huge marketing opportunity and again be overhyped. If you are a middle-aged man who smokes and is somewhat overweight, you don't need a genetic test to tell you that you are at risk of heart disease. But if genotyping becomes the norm, I can see an explosion in the market for diet books, nutritional supplements and exercise programs designed for people with specific genetic profiles. I have a nightmare that people will choose which restaurant to eat at according to their genotype. It will be a mess, it will be overdone, but there will be some germs of truth in what the tests are saying.

What I think is much more important and much more realistic in the timescale of a decade or so is the prospect that we will find new drug targets for diseases that we currently find very difficult to treat. For example, Mike Stratton's cancer group at the Sanger Centre is screening tumors to see how they differ genetically from normal tissue. In many cases it may be easier to kill a cell than to cure it.

Genome information should help to reveal targets on the tumor cell so that drugs can seek them out and destroy the tumor cells selectively, leading to fewer side-effects and higher cure rates than in conventional chemotherapy and radiotherapy. It's likely that in ten or twenty years' time many more cancers will be treatable than today.

When people started talking about cures for genetic disease, twenty or thirty years ago, they generally spoke in terms of gene therapy: replacing a bad allele with a good one, or genetically transforming cells to produce useful products, such as the growth factors that can help damaged brain cells to regenerate. Laboratory research has laid most of the groundwork for this approach, but successful gene therapy is turning out to be a more elusive goal than was hoped. The best chances of success are in diseases where the cells you want to treat are accessible, such as leukemias or immune deficiency diseases where you can take blood or bone marrow cells out, treat them and put them back: French doctors for the first time successfully treated two babies with severe combined immunodeficiency disease using this technique in 2000. Trials have also been under way for some time in cystic fibrosis, where the cells that need a working gene are in the membranes that line the lungs, theoretically reachable by using an inhaler. So far such treatments have not led to long-term improvements, an indication of just how hard it's going to be to tackle in this way diseases of less accessible tissues, such as the brain and nervous system. Getting genes engineered and delivered and turned on and off properly calls for much greater understanding than we currently possess of how the system works.

But it would be wholly misleading to suggest that sequencing the genome has been a waste of time because gene therapy hasn't worked immediately. This is one area where the hype has been overdone—which is understandable when patients are desperate for cures. But in the long term this should be seen as no more than a momentary setback. There was no more reason why gene therapy should deliver instant cures than any other form of experimental medicine; it is just as promising, though just as doubtful in the early stages of its development, as organ transplants, for example.

Knowledge of the genome could eventually allow parents to endow their children artificially with genes for 'desirable' characteristics such as intelligence, beauty and so on, producing so-called 'designer babies.' For the moment this is an implausible scenario, for a variety of reasons. The route from picking your ideal baby from a catalogue to a successful birth, never mind a healthy child or adult with the chosen characteristics, is long and full of uncertainties. And, as we've seen, such characteristics are likely to depend on sets of genes working in concert in ways we have scarcely begun to understand. So, even if one could overcome the considerable technological obstacles to transforming an embryo, the result might well fail to meet the parents' expectations, with horrendous consequences for the unfortunate offspring (not to mention lawsuits galore). For the moment it's much better to wait for the delightful surprise of producing a new and unique individual by the conventional method—and more practical. The genes are only the start of a person; the environment, and particularly parenting, are immensely important, and in general it's not at all good for children if their parents have specific expectations of them. However, in a generation or two, with advances in knowledge, parents may genuinely have these options and will have to decide.

Negative selection, on the other hand, is already going on. Genetic screening in pregnancy, available now for several years, gives parents at risk of producing a child with a genetic disease such as muscular dystrophy the option of a termination if the diagnosis is positive. Some clinics now offer pre-implantation diagnosis, screening very early embryos produced by in vitro fertilization and implanting only those that have a normal gene. There is a narrow ethical line between this procedure and introducing a healthy gene into an embryo that lacks it—but the latter is currently banned under U.K.

legislation that outlaws germline gene therapy, in which the treatment will not be restricted to the individual but will also be carried down through future generations. This is wise, for apart from ethical considerations, our ignorance about the potential ramifications is too great. Whether that decision will later be reversed in the light of increasing knowledge will be a matter for democratic debate.

These considerations relate to more than just the practicalities. They also raise a whole series of important ethical questions: the rights of the unborn and the rights of parents; the concept of genetically 'normal' and the concept of genetically 'better.' The first half of the twentieth century saw, in both Europe and North America, the horrors of eugenic movements in which individuals decreed genetically 'defective' were forcibly prevented from reproducing or, as in the 'final solution' of the Nazi regime, murdered. Most of us recoil from these events, but there is no escaping the fact of our new powers, and the need to exercise them responsibly. Some would prefer that we deny them by abstaining totally from intervention. Certainly there is a danger that by narrowing too far the boundaries of 'normality' we could deny life to people who should rightfully enjoy it. On the other hand, there are already cases of children bringing lawsuits against their parents for 'wrongful life'— allowing them to be born to lead disadvantaged lives. We cannot evade the need for a balance of rights. In any event, it is essential that, once born, all human beings are treated equally regardless of genetic endowment.

What is most important is that the genome is a key step towards the molecular anatomy of the human body. We are right at the beginning, not the end; we don't know what most of the genes look like, or when or where they're expressed. The genome alone doesn't tell you any of these things. Nevertheless, the information is there as a resource and a toolkit to which people will come back again and again as they build up knowledge of the complete structure of the body from the foundation. The next step is to discover all the genes: to figure out what the genome is coding for, where the genes are and particularly where all the control signals are. Because the coding regions account for only 2 percent or less of the human genome, they are much harder to find than in more gene-dense organisms such as the worm, which has a coding density of about 30 percent. Comparing the human sequence with those of other species such as the mouse or the zebrafish is going to be part of the way forward. Although we have diverged from other vertebrates during evolution, natural selection will have ensured that the coding and control regions that are essential to make a viable animal will have been conserved. So looking for matches between genomes is a good gene-hunting technique that can help to fill in the gaps left by automatic gene prediction programs and matches to cDNAs.

Once we've found the genes, we need to work out what proteins they produce, and to understand their time and place of expression. Investigation in all of these areas is going forward at a great pace. None of these jobs is finite—it's quite unlike sequencing the human genome, because every time you do a gene expression experiment you're going to get a different answer, depending on the conditions. You could in principle set up a factory and try to collect a massive set of data, but it's most useful if people who are working on particular mechanisms of the body each carry out their own studies on the tissues that interest them.

Then there's the idea of collecting all the proteins, looking at all the interactions between them—'proteomics' is the new hot area for both academic labs and private companies. You can portray it as a Tower of Babel—'You can't possibly understand all that!'—but because people work on subsystems, we will gradually fit together these pieces of the mechanism. Richard Dawkins's lovely phrase the 'blind watchmaker' is exactly right: we are finding all these little pieces of clockwork that have been put together in the most irrational way to make the whole thing work. And the neat thing about the pieces of mechanism is that many of them are the same in the human as they are in the worm and the fruit fly. Some of the most fundamental mechanisms, such as the cell death pathway that removes unwanted cells, were first studied in the worm, and some of the same genes control programmed cell death in the human.

Somewhere in the genome will be the answer to what makes us different from all the other species—what makes us human. But it's very unlikely to be as simple as having a gene or two that chimps don't have. We will need to know much more about how the whole system acts in concert before we truly understand ourselves.

At Francis Collins's suggestion we concluded our paper on the draft genome with an ironic echo of Watson and Crick's famous understatement in their 1953 announcement of the structure of DNA. 'It has not escaped our notice,' we wrote, 'that the more we learn about the human genome, the more there is to explore.' By making the genome sequence freely available, we made sure that the number of explorers would be unlimited.

8 OUR GENOME

IT WAS 9 JUNE 2001 IN THE SMITHSONIAN MUSEUM OF NATURAL History in Washington DC. That weekend there was a meeting of the Genetic Alliance, a gathering of interest groups dedicated to supporting their genetically disadvantaged members. Francis Collins and I had just begun the proceedings by giving a joint presentation of the nuts and bolts of the Human Genome Project. At the end of the talk Francis picked up his guitar, as he often does at such events, and sang about the human genome, words he'd composed that morning to the tune of a folk song called 'For all the good people'. The chorus went:

> This is a song for all the good people
> All the good people whose genome we celebrate
> This is a song for all the good people
> We're joined together by this common thread.

I was delighted that he'd used the very phrase that we'd hit on for the title of this book, so I compared notes to see if we'd mentioned it already. No, it had just come to him out of the blue. It must be an icon.

Like Francis, most of the characters in this drama are today continuing in their accustomed roles. For my part, I left the scene as
intended, only to find myself not backstage but in another theatre
and invited to keep performing. Media exposure and a knighthood
have handed me a small platform. The knighthood is a tremendous
honor, of course, but one to which I felt hardly entitled, given my
limited contribution to the actual sequencing operation; after some
hesitation I accepted with gratitude on the basis that it was a richly
deserved recognition for the achievements of the Sanger Centre as a
whole. Now, having got my invitation, have I got anything to say?
As Tom Lehrer wisely remarked, 'If a person can't communicate,
the very least he can do is to shut up.' But it is a commonplace that
science can do with more rather than less communication, so it's
incumbent on me to have a go for a while and see whether I can
make a useful contribution. This chapter is one attempt to find out.

My other colleagues in the Human Genome Project remain
engaged in saner pursuits closer to their calling, taking forward the
human and other genomes while diversifying into interpretation
and application. They and the public databases are the guarantors of
the human genome. The large public labs are continuing to collaborate on new activities; the international network is thriving, with a
meeting in Beijing in 2001 hosted by Huanming Yang, and many
other private and public genomic initiatives under way.

The founders of Celera, meanwhile, have succeeded in becoming
wealthy. However, in January 2002, its shares down to a tenth of
their peak value, the company announced that Craig Venter was
stepping down as President: Celera's parent company, Applera, now
envisages its future in drug discovery rather than sequencing and
databases. The earlier claim to be the 'definitive source' of genomic
information seems to have been abandoned—Celera now describes
itself as a 'leading provider' of such information. But the struggle to
keep genomic data free continued to the end, as we saw with the
mouse genome, and indeed may never be over.

So, what would have been wrong with leaving it to a company? Just that, to the extent that the data are fundamental and important, they should be available to all on equal terms, not to the wealthy few. In addition, just as I found back at the beginning with ABI software, Celera tried to broaden its hold over the data. In signing up to Celera databases, academics had to agree to download what they needed for their own use but not to redistribute the data. This was essential, of course, to protect the company's business, but it meant that the normal exchanges of bioinformatics were inhibited, could take place only through the company's database, and were restricted to subscribers. How many biologists really think that this is a good way to run their research? Not many, I suspect, which is why there is general support for continued public sequencing. But they should be wary of inadvertently supporting a slide towards monopoly.

Or, in a few cases, advertently. Some scientists have written articles that uncritically reproduce Celera's claims to sequence much faster and much more cheaply than anyone else. In so doing they acted as volunteer advertisers, for as we've seen these statements are not supported by the facts. The whole affair has been a remarkable example of the Emperor's New Clothes. But so long as the funding agencies continue to support public sequencing, no great harm will have been done.

In common with most of the publicity about the human genome, I suppose this book is really premature. The project is by no means over, despite all the fuss, and to be writing an account already is perhaps presumptuous. But the draft sequence has been published, and the funding and expertise are in place for completion by 2003. So maybe it's not so bad to take stock now.

First of all, where does this achievement—draft and finished—really stand in the overall scheme of things? Is it a Big Idea, or just an episode?

The sequencing of the human genome is not in itself one of the big ideas, but it is a milestone embedded in the big idea of molecular biology. Molecular biology as a whole is about understanding the parts and processes of life in sufficiently complete detail, which in practice means at the atomic level, to predict the effect of alterations. In so doing it is continuing the process, begun by organic chemistry, of bridging the once-perceived chasm between living and non-living matter. At one time the very molecules of life were thought to be special, requiring the intervention of a 'vital force' for their synthesis. So when in 1828 Friedrich Wöhler synthesized the organic substance urea from inorganic materials it was the beginning of a revolution in thinking, a first demonstration that the molecules of life are not sacrosanct in themselves. But for long after that the elaborate organization of living things remained daunting and mysterious, and left plenty of room for vitalism as a respectable concept. It has only been through the triumphs of molecular biology (really just another term for the chemistry of life) in the second half of the twentieth century that we have begun to see our own bodies as exquisitely comprehensible machinery. It would be unwise to predict that this understanding will ever be complete, but there will be a convergence for practical purposes.

An interesting feature of the current stage of knowledge is that we are recognizing that we cannot necessarily distill our comprehension into a simple and elegant theory, as Darwin distilled the theory of evolution from his observations of Galapagos finches and domestic pigeons, but that we can describe it and model it. An important element in the origin of genomics was the willingness to take that step, to say that we really do have to read all this sequence, find all these genes, if we are going to make a model that works. One can argue about whether having a model amounts to real understanding or not. It's interesting that even mathematics has taken this step for some proofs, as in the case of the four-color map theorem. This asserts that any map, no matter how complex, can be drawn in just four colors such that no two areas of the same color share a boundary. One part of the proof of this theorem requires a computer to search through a large set of possibilities: no elegant analytical solution to this part has yet been found. Biologists need not feel uncomfortable that they have to deal similarly with the inventions of the blind watchmaker of evolution. I think that we can reasonably equate prediction with understanding; but, like the mathematicians, biologists have to use computers to sift the information.

A key discovery of molecular biology was that DNA is the hereditary material, that it encodes the instructions to make each living organism and that it is possible to read out the code into a computer. The central task of the Human Genome Project has been simply to read that code as accurately as possible. Of course, we want to understand it fully as well, but that will be a much longer process involving the whole community of biologists.

Apart from its importance as a foundation for the future, our ability to read out the sequence of our own genome has the makings of a philosophical paradox. Can an intelligent being comprehend the instructions to make itself? So far we understand the code so imperfectly that we aren't yet facing that paradox, but there's every reason to anticipate that we shall do so in the not-too-distant future. To put this in perspective, the next big idea will probably be the understanding of the mind—or, more precisely, how the brain computes the mind. It is at this point that the real philosophical paradox of the intelligent organism will arise. Perhaps that is the reason why some say it will never be reached. But then, no one person fully understands the working of a large aircraft or a complex computer. We shall get there by modelling, using computers and understanding one piece at a time. Our understanding of the brain will parallel our modelling of it, but in this case the model, if it works, really will be a brain in its own right.

Not everyone finds such a prospect comfortable. Indeed, there is quite a widespread feeling that science has already gone too far—that it has outstripped human ability to comprehend or control it, and that it must be shackled firmly to social needs, with no opportunity for discoveries that may cause trouble. For example, the difficult dilemmas of prenatal choice, outlined in the previous chapter, may be felt to be the result of science unnecessarily opening Pandora's Box.

However, we can't have it both ways. Scientific method, driven by a desire to explore the unknown, has proved remarkably effective at increasing human understanding of the natural world. It has played a huge role in the development of human culture, and for centuries has contributed in the most fundamental way to philosophy. Awareness of the facts of the universe around me has a huge influence on the way I think about the human situation. Through science, humanity is pushing back the boundaries of ignorance, so that the big questions ('Why are we here?' or 'What is good and what is bad?') can be posed more precisely, framed in a larger area of knowledge. I don't find it unsatisfying that we have not yet arrived at absolute answers to these big questions, and indeed may never do so. There is so much still to find out that they can wait; for the present, what we've already found out gives us plenty to think about.

The past century has seen a split between the sciences and the humanities. Many no longer perceive science as culture. I think much of this attitude is due to science becoming more and more equated with *technology*, to the extent that in many quarters its sole purpose is seen as technological development. This has become an integral part of its funding structure, so that scientists are encouraged to exploit their discoveries commercially, regardless of social consequences. Worse, the development and exploitation are driven by short-term profit, pitting individuals, companies and nations in competition with one another in a frenzied rush for next quarter's bottom line.

But it is not Pandora's Box that science opens; it is, rather, a treasure chest. We, humanity, can choose whether or not to take out the discoveries and use them, and for what purpose. Leaving the chest closed is not an option. Apart from anything else, if some of us don't open the chest visibly and benignly, others will do so secretly and perhaps malignly. Most of the treasures within can be used for either good or ill, but until we've seen them how can we tell? So we must never hold back from exploring. This is our joy and our future as the human race. This is not, of course, true just for scientists. We all without exception know curiosity, the excitement of discovery and understanding, the thrill of doing new things. Knowledge itself is a good: more is always better. But the application of knowledge is a choice, and we have individual and collective responsibility for that choice. Our economic structures are getting in the way of responsible choice, because they drive us to equate discoveries with technology, and to assume that exploitation of knowledge is inevitable. There is no easy solution, but the first step is to recognize the problem.

The history of the human genome sequence illustrates how important it is that we keep clear this distinction between discovery and technology—between science and its commercial application. A high priority for the Human Genome Project has been to keep the genome data completely free. But why is this so important? Why can't it be owned, and why shouldn't at least some restriction on redistribution be allowable, so that the originator of the information can be protected from competition? Many commentators have seen the public project as unreasonable in pressing for the absence of even the latter clause. This is an important matter, because in our society information is more and more the stuff of wealth creation, and the genome is just one example of information that has to be managed equitably.

My first response is that the genome sequence is a discovery, not an invention. Like a mountain or a stream, it is a natural object that was here, if not before we were, at least before we were aware of its existence. I am one of those who feel that the earth is a common good, and is better not owned by anyone, though almost all of us fence off small parts of it for particular uses. Daphne and I 'own' half an acre of land, and though you are very welcome to come and visit I don't expect you to tear up our crops or capture the birds that perch in its trees. But we really only have it on lease; I hope it won't happen, but if there is a democratic decision to build a road through our garden then it will be taken over. A majority of people agree that there should be large tracts of wild places kept aside that belong to no one person but where any of us can go. This compromise works on the basis that we don't have to go everywhere all the time, but that if an area proves important because it's especially scenic, say, or is home to a rare species, then it can be protected as a public good. Of course, we shall always continue to argue about the balance between private land and public land, and the uses to which both can be put.

The human genome is a much more extreme case of the same thing. We all carry our personal copies of it around with us, and every part of it is unique. You can't ever say that you own a gene, because then you'd be owning one of my genes as well. And you can't say, 'Well, we can share the genes between us,' because we both need all our genes. A patent, of course, does not give you literal ownership of a gene, but it does specifically give you the right to prevent others from using that gene for any commercial purpose. It seems to me that your fencing off of a gene should be confined strictly to an application that you are working on—to an inventive step. I, or someone else, may want to work on an alternative application, and so need to have access to the gene as well. I can't go away and invent a human gene. So all the discovered part of genes—the sequence, the functions, everything—needs to be kept pre-competitive and free of property rights. After all, part of the point of the patent system is to stimulate competition. Anyone who wants to make a better mousetrap has to invent around existing mousetrap patents. You can't invent around a discovery; you can only invent around other inventions. As we saw in the last chapter, the most valuable applications for a gene are often far down the line from the first, easy, ones, so this is not just a matter of principle but has extremely important consequences.

In March 2000 Human Genome Sciences, the company set up alongside TIGR in 1992 (TIGR severed the connection five years later), announced that it had been granted a patent on a gene called CCR5, which encodes a receptor on the cell surface. When the company first applied for the patent it did not know what the receptor did. While the patent was pending, a group of publicly funded researchers within the National Institutes of Health had discovered that some people with defects in this gene were resistant to infection with the AIDS virus HIV. CCR5, in other words, appears to be one of the gateways the virus uses to enter cells. As soon as they heard about the discovery, Human Genome Sciences was able to confirm the role of CCR5 through experiments and have the patent issued. It thereby claimed ownership of the rights to the use of the gene for any purpose. The company has now sold licenses to several pharmaceutical companies to develop drugs and vaccines based on this knowledge. But who made the inventive step? The company that made a lucky match to a randomly selected EST? Or the publicly funded researchers who identified that in people resistant to HIV the gene was defective? William Haseltine at Human Genome Sciences argues that patents stimulate progress in medical research, and the CCR5 patent may well lead to a new drug or vaccine against HIV. But a survey of researchers in United States university labs found that many had been deterred from working on particular gene targets because of the fear that they might have to pay large license fees to companies—or risk being sued.

The guidelines on patenting genes in the United States have recently been clarified to give a somewhat tighter definition of utility—the use must be 'substantial, specific and credible'—and rule out the most speculative applications, but they still allow sequences to be patented on such grounds as that they can be used as probes for a known disease gene, for example. The European Patent Directive, approved by the European Parliament in 1998, accepts that a sequence or partial sequence of a gene is eligible for a 'composition of matter' patent once it has been replicated outside the human body—say, copied in bacteria as we do for sequencing. This argument has always seemed to me absurd. The essence of a gene is the information—the sequence—and copying it into another format makes no difference. It is as though I took a hardback book that you had written, published it in paperback, and called it mine because the binding is different.

At the moment, the practice of granting biological patents is not heeding the distinction between discovery and invention, largely because of the immaturity of the field. Twenty years ago patents in biology were almost unknown, and it took a major investment to find a single gene. Now there is a genome gold rush as the industrial potential of the new developments is recognized and indeed overblown, and finding genes can be a matter of spending five minutes at a computer keyboard. The number of applications for gene patents on humans and other organisms has passed half a million, and several thousand such patents have been granted. But the issue of gene patents remains complex and confused. The United States Patent and Trademark Office accepts that a gene discovery is patentable, and until the recent changes would grant patents even on partial gene fragments whose only claimed utility was as 'a gene probe.' The European Patent Office was more doubtful about gene patents until the European Union issued its 1998 patent directive, which explicitly permitted the patenting of gene sequences. However, several EU member states, particularly France, are opposed to gene patenting and have challenged the directive. Meanwhile other European countries, of which the U.K. is one, feel that they must encourage a more liberal line on patenting so that their biotechnology industries can remain competitive with those in the United States.

I realized long ago that trying to reach an equitable solution using

269

moral or even legal arguments was doomed to failure, and that the best way to prevent the sequence being carved up by private interests was to put it into the public domain so that, in patent office jargon, as much as possible became 'prior art' and therefore unpatentable by others. And I think we in the international sequencing consortium succeeded in doing that as far as the raw sequence was concerned. Now, by making the annotated sequence available through Ensembl and other genome browsers, we are pushing the bar still higher by putting information about gene function in the public domain.

But the bar is not high enough yet. For example, although would-be patentees must demonstrate a credible use for a gene sequence before they can patent it, it is generally assumed that the patent then applies to all uses for that sequence, not just the one cited in the application. It will take a legal challenge to establish whether this is correct, but in the meantime it is a disincentive to further research on the sequence concerned. But it seems unlikely to me that patent laws combined with untrammelled market forces are going to lead to a resolution that is in the best interests of further research, or of human health and well-being. Surely there is a case for governments to regulate what's going on? I would prefer to see patents restricted to specific tests and drugs; but if this is too idealistic, a pragmatic step would be to make gene patents subject to compulsory licensing with affordable royalty payments, so that no company could monopolize part of the genome and charge exorbitant fees for its use.

In due course all this hyperactivity will settle down, but for the moment there is the atmosphere of a lottery as everyone tries to buy a winning ticket. As well as patenting human genes, companies are staking claims on natural products used for centuries in the developing world, giving rise to justified accusations of 'biopiracy.' For example, Western companies have been granted over 100 patents on various uses of the neem tree, whose seeds, twigs and leaves have been used in India for centuries in health care and agriculture. Anti-biopiracy campaigners won a significant victory in May 2000, when the European Patent Office revoked a six-year-old patent held by the United States Department of Agriculture and the agrochemical company W. R. Grace for the use of extracts of neem seeds as a fungicide. The Patent Office accepted that this use was not novel and involved no invention—but it took a protracted legal battle to win the case.

Understandably appalled by what is going on, some have proposed to draw a patent line between life and non-life. While agreeing with the concerns, and with the urgent need for a value other than a commercial one to be placed on living things, I think there is no case for this particular line. Because the chasm that previously existed between the biological and the chemical is being filled in, such a distinction will not be sustainable. Yet surely we should not be patenting whole life forms, such as transgenic mice or cotton plants? True, but not just because they are life forms: a sounder reason is that we have not invented the organism, only the specific change that makes them susceptible to cancer (in the case of the mouse) or resistant to pests (in the case of the cotton). Very probably at some point we shall invent new life forms, but that's for the future. All that should be patented now are the modifications that are being made, which are truly inventive steps.

You may feel that this line of argument implies a disrespect for life. But I think that any such concern is a result of the increasing tendency of our society to define the value of everything in financial terms, and to assume that any form of exploitation is justifiable if it shows a profit. This is a separate and larger issue, to which we shall return later.

A second response to the question of why we need to keep sequence information free is the need for it to be easily exchanged between researchers, which we have already touched on. The future of biology is strongly bound up with bioinformatics: that is, the field of research that collects biological data of all kinds, and tries to make sense of the data as a whole, and to make predictions. This activity is essential to give wide access to the data, and to complement and connect with the work of the experimental biologists. Analysis of sequence is one of the foundations of bioinformatics, because the data are stored in computers anyway, and self-evidently need analysis. Computers also analyze the three-dimensional structure of proteins, and tackle the challenge of predicting how that structure will emerge from the folding of a chain of amino acids. Then there are the interactions among the molecules, which build up the actual shape of cells and organisms. Most difficult of all, perhaps, is to understand how all this is controlled, a field which is only just beginning to be explored.

Putting all this together, we can say that the ultimate aim of biology is to compute an organism from its genes (being mindful of the role of the environment and random factors in development): to understand all the processes so well that we can predict the whole from the sequence, just as the mechanisms of our bodies do. To do this completely is a far-off dream, but large sections of the problem will be solved in the coming decades, and this is the most important single reason for acquiring the sequence in the first place.

In order to move forward with this enterprise, which is not only fascinating science but will translate into medical advances as it goes forward, the basic data need to be freely available to everyone to interpret, change and pass on, just as in the open source model for software. The situation is too complex for this to be done piecemeal, with limited amounts of data being let out at a time and a single enterprise always holding the key. Although we concluded in 2000 that we were unable to enforce free redistribution of data through some kind of 'copyleft' license, since the data were not ours to make rules about, we could insist that our own right to make them initially available in this way was not compromised by any agreement we might make. That is why the talks of December 1999 came to nothing.

I expect that the public data will continue to be the ultimate resource, because as well as being available to all they will be continually enriched by the free interactions surrounding them. The only danger is that, being free, the data can be picked up and presented as belonging to someone else, just like my paperback copy of your book in the absence of copyright law. This could allow a company bent on gaining a monopoly position to add public data to its own and then claim that, as its product is superior, publicly funded sequencing and analysis are unnecessary. The defense is for us all, but scientists in particular, to be aware of what's going on and not be gulled by the claims to greater efficiency made by private enterprise, which on close examination usually come down to presentation. Make no mistake: publicly funded science is extremely efficient because it is ruthlessly competitive, as we have seen in the case of the HGP. The success of the Sanger Centre and the other big genome labs has shown that size is not an issue: the often-heard notion that only industry can handle large-scale science is incorrect.

Another potential threat to academic freedom is the courting of corporate donors by cash-strapped universities. When Nottingham University accepted nearly £4 million from British American Tobacco to fund an 'International Centre for Corporate Social Responsibility', 85 percent of respondents to a poll in the *British Medical Journal* condemned the university for taking the money. The editor of the *BMJ*, Richard Smith, resigned from the post he held at the university as professor of medical journalism. But all universities now hold contracts with industrial sponsors out of sheer necessity; the question is to what extent the sponsors thereby gain control over what is and what is not discovered. Contracts typically protect the researcher's right to publish. But once a department is dependent for jobs and research funding on a certain source of revenue, what happens at renewal time? The pressure to be accommodating is then huge.

I don't want to sound hysterical about this. The art of running a great university is to extract funds from all possible sources and to balance them against one another so as to assure one's independence. Astronomy was founded by people whose patrons thought they were studying astrology on their behalf. But it's necessary not to be too greedy, to work sufficiently within one's limitations that one can afford to say no to excessive pressure.

The pressure to exploit comes not only from companies but from government and charitable funders, both of which are anxious on the one hand to use their finite resources as sparingly as possible and on the other to provide a justification for their activities. In the U.K., for example, since the mid-1990s the research councils have been specifically charged with supporting work that will contribute to 'wealth creation and the quality of life.'

The commercial and competitive pressures on academics today are alarming. And if academics are not independent, who will be society's impartial experts? To maintain the system, scientists need constantly to reaffirm collectively that open communication works and is necessary for their research to prosper.

Am I claiming too much for the co-operativeness of most scientists? Good science is a free-market and freelance sort of enterprise. Restrictions and planning are anathema to it, and anarchy is an essential part of the process. Anyone can challenge anything, and you're only as good as your last five years. (In business and politics the credibility span is pretty short also, and maybe society suffers from the resulting short-termism.) As far as funding goes, many scientists are adapted like desert plants, putting out long roots in all directions; when the rain falls they suck it up fast, and flower. Furthermore, the origins of science are in industry as well as philosophy, an activity of artisans and industrialists as well as intellectuals. I should beware the lessons of the Greek empire, in which thought was elevated to such a degree that it became separated from action, which was the preserve of the slaves. So the philosophy of the Greeks lost touch with reality, and in the end they had to submit to the newly pragmatic Romans.

So this is not a call for science to be run by committee—that would be counterproductive. It's purely about the ethic of science, which recognizes the commonality of the ever-growing body of knowledge and the need for it to be freely available to all, for any purpose.

The events described in this book are part of a much larger picture. It was not, as I fondly imagined at the beginning, simply a matter of sequencing the human genome and making the data available. This was naive. I'd thought of the Human Genome Project as being an uncluttered and altruistic activity, but found instead that others viewed it as a stepping stone on the route to commercial profit or political power. I was forced to realize that in our society one can get into trouble for giving away something that can make money. I began to notice parallel tragedies unfolding, stories in which I was not personally involved but which drew me into discussion of areas outside my own small expertise.

After the June announcement, David Bryer, then director of Oxfam, wrote to ask if I would be interested in a meeting. He enclosed a copy of their recent report to the government on globalization, which echoed my own thoughts so accurately that I was stunned. Daphne and I are lifetime supporters of Oxfam, and although I had no idea how I could contribute it seemed a good idea to explore. So one freezing, foggy morning I drove over to Oxford. We talked, and I gave a seminar over lunch. It was very well attended, and evidently the Human Genome Project and the events surrounding it were of great interest to the researchers there. For my part, I found a group of wise and thoughtful people who saw clearly that it was no good drilling artesian wells and distributing food without paying equivalent attention to the long-term causes of poverty. And their thoughts had come inexorably to the subject of world trade and the manner in which the economic practices of the

World Trade Organization were actually increasing the gap between rich and poor in the world rather than narrowing it, which must surely be the only sane course if we wish to continue inhabiting the planet.

In particular, they were about to embark on a new campaign to arrest the savage implementation of the WTO's agreement on trade-related aspects of intellectual property rights (TRIPs). This agreement provides for the universal extension of patent law, with the full protection of inventors' rights for 20 years worldwide. If carried through, it would ensure a large and immediate increase in the cost of medicines in the poorest countries, because they are at present dependent upon buying generic (non-branded) drugs made by companies in India, Egypt, Brazil, and elsewhere, and these sources would be challenged by the patent-holders forthwith. The consequences for the fight against major killers such as AIDS and dysentery would be devastating.

The argument here is different from that over patenting genes. The drugs, at least in the particular applications for which the companies are making them, are legitimate objects of patent law. It is rather the hasty and over-zealous implementation of the TRIPs agreement against those who are unable to defend themselves that Oxfam is criticizing. As in the case of the debt burden of poor countries, the rich may well pause and say to themselves 'Is this ethical, this application of our laws indiscriminately to all? And is it even good business practice, seeing that it will make the world an even less stable place than it is already?'

The WTO defends its position by pointing out correctly that there are built-in safeguards to the TRIPs agreement, by which in case of urgent medical need countries can make special arrangements. The difficulty is that, as always in the pursuit of justice, those who have the money to pay for skilled lawyers are at an advantage. South Africa and Brazil found that they were facing lengthy lawsuits to establish their rights in this regard, and were at the same time being threatened with trade reprisals if they dared to press their case. The power of the rich countries and of the transnational corporations was being used in a bullying and inequitable fashion to achieve ends that benefit them rather than mankind as a whole.

I was happy to be associated with the campaign, for it echoed on a global scale the issues we had fought out over the genome. During 2001 there were extensive discussions of these issues, and a major court victory over the drug companies in South Africa, but these are only preliminary steps; the vested interests that gestated the TRIPs agreement are alive, vigorous, and no doubt regrouping. For patents, successfully defended, mean money.

The big transnational corporations are now more powerful than many governments. Their strength is apparent everywhere we turn, and especially in their collective lobbying in the capitals of rich nations. Maybe we're moving towards a world where national governments, elected or otherwise, no longer count. Or at least they will count only for local affairs, rather as local councils do now: local cooperatives, nothing more. I hope this isn't true, but the warning signs are there.

Will anything else offset the power of companies, and provide some democratic limitation to their ambitions? A likely source of balance is from the non-governmental organizations (NGOs), such as Oxfam. The largest of them are already transnational, too. Like companies, they are controlled by those who invest in them; but unlike companies, their aims are ethical rather than financial. It's strange and uncomfortable for me to think that democracy may one day be practiced through this balance of power, but the signs are that it's a real phenomenon.

We are often told that the enterprise of science pays too little attention to the consequences of its discoveries and the concerns of society. Should scientists see themselves as part of a worldwide NGO, upholding a set of shared values? I think that's exactly the way they used to be: in previous centuries, as wars ebbed and flowed, the intellectuals made their own independent way around the world. And actually that international fellowship is by no means gone, but it's threatened when people try to walk both sides of the line, mingling scientific contribution with profit-making activity. The two do not mix well, which is why there have to be rules to keep them identifiably separate. Science itself, as well as society, will be the poorer if commercial imperatives are allowed to dictate our terms of reference.

The truth is that companies don't have to behave ethically: they can if they want to, but there's no social constraint on their pushing acquisitiveness to the legal limit; or indeed beyond, in that the cost in fines of an action that breaks legal bounds can be estimated and written off against the likely gains. Similarly, the legal constraints on claims made in advertisements and press releases are continually pressed, and sometimes overreached. And in our overly PR-conscious society there is little questioning of a smooth presentation. Half truth that is branded with a recognized name and laminated to cover the cracks is rated more highly than unvarnished fact.

Of course, in the commercial world this is absolutely natural and right. The job of publicly quoted companies is to maximize their profits. They may, and often do, achieve this aim by taking a long-term view, contributing to good causes, treating their workers well, and so forth. But all these things must be justified by the eventual return on the balance sheet. A company that behaves in any other way will be displaced or will be bought out by a more competitive rival. It is both the strength and the limitation of capitalism that this is so. If we collectively want a company to provide a public good we have to specify how it will be delivered. It's no use leaving the company to set the rules.

We in Western society are going through a period of intensifying belief in private ownership, to the detriment of the public good. Individual selfishness is held up as the best way to advance civilization, and through the process of globalization these beliefs are being exported to the world as a whole, making it not only less just but also less safe. For it's not only the companies. As nations, too, we are unable to take sensible collective decisions when the only rules we know for bargaining are those of competitive greed. It cannot be right that the wars currently being waged in some of the poorest countries in the world depend on weapons sold to their armies by the richest, including my own; that we are unnecessarily destroying natural resources for lack of cooperative exploitation; that we are unable to reach accord on global warming because it might slightly reduce the economic growth of the world's richest nation; and that the disparity in healthcare between rich and poor is widening all the time.

And as this story shows, the same greed nearly succeeded in privatizing the human genome, our own code, and indeed remains a threat to it. But the Human Genome Project has achieved its first target, to have the draft human sequence out and available for all to use, and that is a splendid victory. It's moving on fast to its next target—to produce an accurate, finished sequence—and that is secure and nearly complete.

Whatever happens, let nobody be tempted to rewrite history. The struggle over the human genome was necessary, and things would not be the same today had not the public project stood firm.

The discoveries arising from the human sequence mount up all the time, but that isn't really the point. The important thing is that it has entered the fabric of biology, and, like the worm map and sequence before it, will soon no longer be visible as a separate item. That is as it should be. People often talk about the post-genomic era. Wrong. It's merely the post-hype era. The sequence is now a crucial element in a free and open system of biological information that will allow knowledge to increase and benefits to accrue faster than in any other way.

It is our inalienable heritage.

It is humanity's common thread.

NOTES

We consulted the following two books extensively for background on the early history of molecular biology and the Human Genome Project respectively:

Horace Freeland Judson, *The Eighth Day of Creation* (Jonathan Cape, 1979; paperback edition Penguin, 1995).

Robert Cook-Deegan, *The Gene Wars* (Norton, 1994).

These are referred to in the notes below as Judson and Cook-Deegan.

For a lucid account of what the influence of our genes means for our understanding of ourselves, see:

Matt Ridley, *Genome: The Autobiography of a Species in 23 Chapters* (Fourth Estate, 1999).

1 WITH THE WORMS

p. 3 Francis Crick famously shouted: James Watson, *The Double Helix* (Atheneum, 1968).

p. 3 So when the contemporary artist Marc Quinn: I met Marc when the National Portrait Gallery commissioned him to make a portrait of me for an exhibition in 2001–2. The portrait featured my own DNA growing in bacterial clones.

p. 4 You can do a similar experiment: for example, Susan Aldridge gives instructions in the opening chapter of her book *The Thread of Life* (Cambridge University Press, 1996).

p. 5 Sydney had come to Oxford: Judson, p. 231.

p. 6 'We propose to identify every cell in the worm': letter from Sydney Brenner to Max Perutz, cited in Sydney Brenner's foreword to *The Nematode Caenorhabditis elegans*, edited by W. Wood and the community of *C. elegans* researchers (Cold Spring Harbor Laboratory Press, 1988).

p. 6 'Jim Watson said at the time that he wouldn't give me a penny ...': Roger Lewin, 'A worm at the heart of the genome project', *New Scientist*, 25 August 1990, pp. 38–42.

p. 6 one of the main objects of his program: Sydney Brenner, 'The genetics of *Caenorhabditis elegans*', *Genetics*, vol. 77, 1974, pp. 71–94.

p. 7 In the wild, *C. elegans* lives ...: Donald Riddle et al., 'Introduction to *C. elegans*', in *C. Elegans II*, edited by Donald Riddle et al. (Cold Spring Harbor Laboratory Press, 1997).

p. 8 ... skeptics who had said that the worm was so boring: ibid.

p. 9 'desks encouraged time-wasting activities': John White, 'Worm tales', *International Journal of Developmental Biology*, vol. 44, 2000, pp. 39–42.

p. 9 another new arrival: Mike Wilcox, one of my closest friends in the early days; our families spent a lot of time together. Sadly, he died of cancer in 1992.

p. 10 In May 1947 he wrote ...: Max Perutz, 'How it all began', *MRC News*, Winter 1997, p. 1.

p. 10 'Watson's arrival had an electrifying effect ...': Max Perutz, 'How the secret of life was discovered', in *I Wish I'd Made You Angry Earlier* (Oxford University Press, 1998).

p. 11 with the crucial help of an X-ray photograph of DNA taken by Rosalind Franklin: see Anne Sayre, *Rosalind Franklin and DNA* (Norton, 1975).

p. 11 ... published it in *Nature* ...: James Watson and Francis Crick, 'A structure for deoxyribose nucleic acid', *Nature*, vol. 171, 1953, pp. 737–8.

p. 11 Watson and Crick followed up with another: James Watson and Francis Crick, 'Genetical implications of the structure of deoxyribonucleic acid', *Nature*, vol. 171, 1953, pp. 964–7.

p. 13 Fred Sanger was the first to work out the complete sequence of amino acids: F. Sanger and E. O. P. Thompson, 'The amino-acid sequence in the glycyl chain of insulin', *Biochemical Journal*, vol. 53, 1953, pp. 353–74.

p. 13 Perutz and Kendrew had done the seemingly impossible: Max Perutz et al., 'Structure of haemoglobin', *Nature*, vol. 185, 1960, pp. 416–22; John Kendrew et al. 'Structure of myoglobin', *Nature*, vol. 185, 1960, pp. 422–7.

p. 13 Fred Sanger's 1975 invention of a method for reading the sequence of DNA: Fred Sanger et al., 'Nucleotide sequence of bacteriophage phiX174', *Nature*, vol. 265, 1977, pp. 687–95.

pp. 13–14 Altogether nine Nobel prizewinners: James Watson, Francis Crick, Max Perutz, John Kendrew, Fred Sanger, Cesar Milstein, George Köhler, Aaron Klug, John Walker.

p. 14 [Max Perutz] was first interned as an alien: Max Perutz, 'Enemy alien', in *I Wish I'd Made You Angry Earlier* (Oxford University Press, 1998).

p. 14 Francis himself put it best in a letter: letter from Crick to Watson, 13 April 1967, cited by Judson, p. 182.

p. 23 I learned to do a reaction: the tutors on this course were Ed Furshpan and David Potter, neurophysiologists from Harvard Medical School.

p. 24 Sydney put me on to determining the quantity of DNA: John Sulston and Sydney Brenner, 'The DNA of *Caenorhabditis elegans*', *Genetics*, vol. 77, 1974, pp. 95–104.

p. 25 the first [paper] to come out with my name on it: Gerry Rubin and John Sulston, 'Physical linkage of the 5S cistrons to the 18S and 28S ribosomal RNA cistrons in *Saccharomyces cerevisiae*', *Journal of*

Molecular Biology, vol. 79, 1973, pp 521–30.

p. 26 previous researchers: Simon Pickvance and Roger Freedman had worked on the worm embryo in Sydney's lab before me.

p. 28 'She would go through the successive sections . . .': interview with John White, 5 June 2001. See also John White, 'Worm tales', *International Journal of Developmental Biology*, vol. 44, 2000, pp. 39–42.

p. 28 Eventually published in 1986 . . .: John White, Eileen Southgate, Nichol Thomson and Sydney Brenner, 'The structure of the nervous system of *Caenorhabditis elegans*', *Philosophical Transactions of the Royal Society of London B (Biological Sciences)*, vol. 314, 1986, pp. 1–340.

p. 29 At one point Sydney bet John . . .: this and other anecdotes of the time are related in Robert Horvitz and John Sulston, 'Joy of the worm', *Genetics*, vol. 126, 1990, pp. 287–92.

p. 31 We published the post-embryonic lineage . . .: John Sulston and Robert Horvitz, 'Post-embryonic cell lineages of the nematode *Caenorhabditis elegans*', *Developmental Biology*, vol. 56, 1977, pp 110–56.

p. 31 'I knew that if I went to work with John . . .': interview with Judith Kimble, 13 December 2000.

p. 32 The results of the various attempts: Gunther von Ehrenstein and Einhard Schierenberg in Germany and Teddy Kaminuma in Japan were recording embryos on movie film and trying to reconstruct the cell divisions.

p. 34 After a year and a half it was done: John Sulston et al., 'The embryonic cell lineage of the nematode *Caenorhabditis elegans*', *Developmental Biology*, vol. 100, 1983, pp. 64–119.

p. 34 The lineage . . . 'has given single-cell resolution to worm biology': interview with Bob Horvitz, 11 December 2000.

2 ON THE MAP

p. 38 Matt's group had been working for some time . . .: Matt Scott and Amy Weiner, 'Structural relationships among genes that control development: sequence homology between the Antennapedia, Ultrabithorax, and fushi tarazu loci of *Drosophila*', *Proceedings of the*

National Academy of Sciences USA, vol. 81, 1984, pp. 4115–19.

p. 44 'Baconian science': see e.g. John Pickstone, 'Natural history', in *Ways of Knowing* (Manchester University Press, 2000).

p. 46 Christiane Nüsslein-Volhard: she shared the Nobel Prize for Physiology or Medicine in 1995 for her work.

p. 46 'Lab meetings became nothing but progress in mapping': interview with John White, 5 June 2001.

p. 47 Fred's method [of sequencing DNA]: Fred Sanger shared the Nobel Prize for Chemistry in 1980 with Walter Gilbert, who devised an alternative method.

p. 49 'I saw the crystallographic people . . .': interview with Alan Coulson, 5 December 2000.

p. 51 The LMB workshop built one [a scanner]: Frank Mallett designed the instrument and its data collection software.

p. 52 'It was clear that the way we were going to learn about muscle . . .': interview with Bob Waterston, 8 December 2000.

p. 55 *Worm Breeders' Gazette*: begun by Bob Edgar who visited Sydney Brenner's lab in the early days, this publication is still going strong and can be seen at <elegans.swmed.edu/wli/>.

p. 56 1986, the year we published our first paper on the worm map: Alan Coulson, John Sulston, Sydney Brenner and Jon Karn, 'Towards a physical map of the genome of the nematode *Caenorhabditis elegans*', *Proceedings of the National Academy of Sciences USA*, vol. 83, 1986, pp. 7821–5.

p. 56 'There were four of us . . .': interview with Bob Horvitz, 11 December 2000.

p. 57 'We talked about genomes . . .': interview with Maynard Olson, 15 June 2001.

p. 58 One of the first to dare to think on this scale: see Cook-Deegan, ch. 5.

p. 59 'as we analyzed the problems to be solved . . .': Robert Sinsheimer, 'The Santa Cruz Workshop—May 1985', *Genomics*, vol. 5, 1989, pp. 954–6.

p. 60 'the grail of human genetics...': reported by Roger Lewin, 'Proposal to sequence the human genome stirs debate', *Science*, vol. 232, 1986, pp. 1598–1600.

p. 61 'Doesn't one person really have to finish up that last 10 percent...': quoted in Cook-Deegan, ch. 12.

p. 62 'I would only once have the opportunity...': James Watson, *Annual Report 1988 of the Cold Spring Harbor Laboratory*, 1989, quoted in Cook-Deegan, ch. 13.

p. 62 a Japanese program had been set up: the project was chaired by Akiyoshi Wada, a biophysicist who had received part of his training in the US, and involved partnership with seven Japanese high-technology companies.

p. 63 'I wrote a little note...': interview with Sydney Brenner, 12 June 2001.

p. 63 In February 1989 the UK's Department for Education and Science announced an £11 million grant: David Dickson, 'Britain launches genome programme', *Science*, vol. 245, 1989, p. 1657.

p. 66 'It was impressive...': interview with Bob Waterston, 8 December 2000.

p. 67 'In my conversations with Jim...': interview with Bob Horvitz, 11 December 2000.

p. 67 'My impression was that Jim believed...': ibid.

p. 69 'John was the standard bearer...': interview with Aaron Klug, 30 January 2001.

p. 72 I was working with the LMB workshop: once again, Frank Mallett was the engineer supervising the project.

p. 76 'Rick knew the machine's limitations...': interview with Bob Waterston, 8 December 2000.

p. 78 'You'd sit down at the computer...': interview with Rick Wilson, 12 May 2001.

p. 79 'There were huge gaps...': interview with Bob Waterston, 8 December 2000.

3 IN BUSINESS

p. 84 'That was a very attractive part of it . . .': ibid.

p. 85 'We were worried that Bourke would recruit John and Bob . . .': interview with Jim Watson, 8 November 2000.

p. 85 'I thought the fact that you had two countries coming together was great . . .': ibid.

p. 86–7 The main bone of contention between [Watson and Healy] was the patenting of gene sequences: this episode is documented in Cook-Deegan, chs 19 and 20.

p. 88 'virtually any monkey': quoted in Leslie Roberts, 'Genome patent fight erupts', *Science*, vol. 254, 1991, pp. 184–6.

p. 88 Watson eventually agreed . . .: Cook-Deegan, p. 329.

p. 88 The scientific advisory committee . . . was 'unanimous in deploring the decision to seek such patents': statement quoted in full in Roberts, 'Genome patent fight erupts'.

p. 88 Walter Bodmer confirmed . . .: ibid.

p. 88 'It makes a mockery of what most people feel is the right way to do the Genome Project': Berg, quoted in Leslie Roberts, 'NIH gene patents, round two', *Science*, vol. 255, 1992, pp. 912–13.

p. 89 'I was not anti-American . . .': interview with Jim Watson, 8 November 2000.

p. 89 'It turned out I always had an illegal job . . .': ibid.

p. 91 'To spend money on that scale sensibly is not easy!': interview with Bridget Ogilvie, 2 February 2001.

p. 92 'The first idea was that we would each put in £2 million . . .': ibid.

p. 92 a burst of comment about the episode: for example, Robin McKie, 'Scandal of US bid to buy vital UK research', *Observer*, 26 January 1992; Roger Lewin and Gail Vines, 'US company plans to hijack DNA project', *New Scientist*, 1 February 1992.

p. 93 the gene for Huntington's disease, which had been linked to a chromosome almost a decade before: James Gusella et al., 'A polymorphic DNA marker genetically linked to Huntington's disease', *Nature*, vol. 306, 1983, pp. 234–8.

p. 93 'That was the argument—that you would be able to get results . . .': interview with Aaron Klug, 30 January 2001.

p. 96 'two or three million': interview with Bridget Ogilvie, 2 February 2001.

p. 97 'It seemed very clear to me that although [John] eschewed responsibility he was actually taking it on willy nilly': interview with Aaron Klug, 30 January 2001.

p. 98 'They plied me with sherry . . .' interview with Jane Rogers, 9 January 2001.

p. 98 'I knew how to write down what was needed . . .': ibid.

p. 99 'It was the obvious thing to do . . .': interview with Bart Barrell, 14 February 2001.

p. 100 'We tried to jump from the Factor IX gene . . .': interview with David Bentley, 11 May 2001.

p. 101 'I was deeply integrated into the Guy's unit . . .': ibid.

p. 104 'When it was clear that the Wellcome Trust and the MRC were going to create this larger venue . . .': interview with Bob Waterston, 8 December 2000.

p. 105 he left in July 1992 to set up . . . TIGR: this development is documented in Cook-Deegan, chs 19 and 20.

p. 105 This technique had been developed by Paul Schimmel and his colleagues: S. D. Putney, W. C. Herlihy and P. Schimmel, 'A new troponin T and cDNA clones for 13 different muscle proteins, found by shotgun sequencing', *Nature*, vol. 302, 1983, p. 718.

p. 105 Craig . . . wrote to Jim Watson: Cook-Deegan, p. 315.

p. 105 He published his first paper on his EST work: Mark Adams et al., 'Complementary DNA sequencing: expressed sequence tags and the Human Genome Project', *Science*, vol. 252, 1991, pp. 1651–6.

p. 105 In an accompanying news article . . .: Leslie Roberts, 'Gambling on a shortcut to genome sequencing', *Science*, vol. 252, 1991, pp. 1618–19.

p. 107 Our paper came out in early 1992: Bob Waterston et al., 'A survey of expressed genes in *Caenorhabditis elegans*', *Nature Genetics*, vol. 1, 1992, pp. 114–23. The worm EST work in Craig Venter's lab was led by

Dick McCombie and published in the same issue of the journal. Dick moved to Cold Spring Harbor soon afterwards.

p. 107 I was quoted as saying 'Eight or nine per cent is more like it': Roberts, 'Gambling on a shortcut to genome sequencing'.

p. 107 'The extramural genome community . . .': Statement of Craig Venter before the Subcommittee on Energy and Environment, U.S. House of Representatives Committee on Science 17 June 1998, available at <www.house.gov/science/>.

p. 108 Craig . . . became a multi-millionaire almost overnight: Frederick Golden and Michael Lemonick, 'The race is over', *Time*, 3 July 2000.

p. 108 Despite his frequent claims to have 'developed' the use of ESTs . . .: Statement of Craig Venter before the Subcommittee on Energy and Environment, 17 June 1998.

p. 108 Sydney Brenner . . . had argued in favor of cDNA sequencing: Sydney Brenner, 'The human genome: the nature of the enterprise', *Ciba Foundation Symposia*, vol. 149, 1990, p. 6.

p. 109 'we should leave something for our successors to do': interview with Sydney Brenner, 12 June 2001.

p. 109 TIGR did score a first in sequencing . . . *Haemophilus influenzae*: R. D. Fleishmann et al., 'Whole-genome random sequencing and assembly of *Haemophilus influenzae* Rd', *Science*, vol. 269, 1995, pp. 496–512.

p. 109 After . . . Randy Scott read about Craig's EST work . . .: Gary Zweiger, *Transducing the Genome* (McGraw Hill, 2001), p. 71.

p. 110 Five years before, genomic entrepreneurs had found it almost impossible . . .: see Christopher Anderson, 'Genome project goes commercial', *Science*, vol. 259, 1993, pp. 300–2.

4 MEGALOMANIA

p. 116 'I sat down and started playing with the numbers . . .': interview with Bob Waterston, 8 December 2000.

p. 118 Merck funded a massive drive . . .: David Dickson, '"Gene map" plan highlights dispute over public vs. private interests', *Nature*, vol. 371, 1994, pp. 365–6.

p. 121 The first whole-genome physical map...: Daniel Cohen, Ilya Chumakov and Jean Weissenbach, 'A first-generation physical map of the human genome', *Nature*, vol. 366, 1993, pp. 698–701.

p. 121 At the same time his colleague Jean Weissenbach: G. Gyapay et al., 'The 1993–1994 Généthon human genetic linkage map', *Nature Genetics*, vol. 7, June 1994, pp. 246–339.

p. 122 New bacterial cloning methods: P1 artificial chromosomes (PACs) were developed by Pieter de Jong at the Roswell Park Cancer Institute, Buffalo, New York; bacterial artificial chromosomes (BACs) by Mel Simon at the California Institute of Technology in Pasadena, California.

p. 122 'In the end they more or less threw up their hands': interview with Bob Waterston, 8 December 2000.

p. 125 Bridget Ogilvie and her colleagues at the Wellcome Trust were horrified: interview with Bridget Ogilvie, 2 February 2001.

p. 126 'I didn't quite say so, but I thought that was their responsibility!': interview with Bob Waterston, 8 December 2000. Clone supply was manageable for the Sanger Centre, because we already had sufficient in-house resources and expertise.

p. 127 In an interview about our plan: Eliot Marshall, 'A strategy for sequencing the genome 5 years early', *Science*, vol. 267, 1995, pp. 783–4.

p. 127 The debate continued at the May 1995 Cold Spring Harbor genome meeting: Eliot Marshall, 'Emphasis turns from mapping to large-scale sequencing', *Science*, vol. 268, 1995, pp. 1270–1.

p. 128 Maynard backed us: see also Maynard Olson, 'A time to sequence', *Science*, vol. 270, 1995, pp. 394–6.

p. 128 the first attempt at a genetic map of the human genome: Helen Donis Keller et al., 'A genetic linkage map of the human genome', *Cell*, vol. 51, 1987, pp. 319–37.

p. 129 'We came at it from the point of view...': interview with Eric Lander, 19 May 2001.

p. 129 'They were out of their minds...': ibid.

p. 130 'Back in 1995 I got myself in a lot of trouble...': ibid.

p. 130 'Our thesis is...': John Sulston and the Sanger Centre Board of

Management, 'The Sanger Centre 1995–2002', proposal to MRC and Wellcome Trust, 1995.

p. 131 'In such a climate . . .': Letter from John Sulston to Diana Dunstan, MRC, 5 June 1995.

p. 133 'It was a question of finance . . .': email from Diana Dunstan to John Sulston, 7 August 2001.

p. 134 'The ideas of Wellcome . . .': email from Dai Rees to John Sulston, 13 August 2001.

p. 135 'In the mid-1990s . . .': interview with Mike Dexter, 20 April 2001.

p. 135 'We [the Wellcome Trust] wanted to take a risk': interview with Michael Morgan, 25 April 2001.

p. 137 'They don't want to see their opportunities cut off . . .': email from Bob Waterston to John Sulston, 16 December 1995.

p. 137 When Francis Collins announced funding: National Center for Human Genome Research, press release, 'Pilot study explores feasibility of sequencing human DNA', April 1996.

p. 138 'Some people looking back in history . . .': interview with Francis Collins, 11 December 2000.

p. 138 As *Science* reported: Eliot Marshall and Elizabeth Pennisi, 'NIH launches the final push to sequence the genome', *Science*, vol. 272, 1996, pp. 188–9.

p. 139 'It would have been much better to have accepted John and Bob's idea at the time': interview with Jim Watson, 8 November 2000.

p. 140 One gene, BRCA1, had been located by Mary-Claire King . . .: J. M. Hall et al., 'Linkage of early-onset familial breast cancer to chromosome 17q21', *Science*, vol. 250, 1990, pp. 1684–9.

p. 140 In the summer of 1994 they located the gene . . .: R. Wooster et al., 'Localisation of a breast cancer susceptibility gene (BRCA2) to chromosome 13q12-13 by genetic linkage analysis', *Science*, vol. 265, 1994, pp. 2088–90.

p. 140 'I was concerned about what would happen . . .': interview with Mike Stratton, 23 February 2001.

p. 141 'We discussed it . . .': ibid.

p. 141 the ICR paper came out in *Nature*: R. Wooster et al., 'Identification of the breast cancer susceptibility gene BRCA2', *Nature*, vol. 378, 1995, pp. 789–92.

p. 142 'The Ashkenazi BRCA2 mutation was in our original paper ...': interview with Mike Stratton, 23 February 2001.

p. 142 in its 1996 paper: Tavtigian et al., 'The complete BRCA2 gene and mutations in chromosome 13q-linked kindreds', *Nature Genetics*, vol. 12, 1996, pp. 333–7.

p. 145 data release ... is the issue with which the Bermuda meeting is most associated: for example, see David Bentley, 'Genomic sequence information should be released immediately and freely in the public domain', *Science*, vol. 274, 1996, pp. 533–4; Mark Guyer, 'Statement on the rapid release of genomic DNA sequence', *Genome Research*, vol. 8, 1998, p. 413.

p. 146 'It was crucial that people from the funding agencies ...': interview with Michael Morgan, 25 April 2001.

p. 146 'A lot of the smaller countries did not trust the US': ibid.

p. 146 we won the acceptance of most (though not all) of the genome sequencing community: see e.g. Mark Adams and Craig Venter, 'Should non-peer-reviewed raw DNA sequence data release be forced on the scientific community?', *Science*, vol. 274, 1996, pp. 534–6.

p. 148 an article in *Science*: André Goffeau et al., 'Life with 6000 genes', *Science*, vol. 274, 1996, pp. 546–67.

5 RIVALS

p. 149 'the definitive source of genomic and associated medical information': press release issued by the Perkin-Elmer corporation and TIGR, 9 May 1998.

p. 150 'I wondered why Craig didn't call me himself': interview with Jim Watson, 15 December 2000.

p. 150 'He told me that Craig was going to make a major announcement': interview with Michael Morgan, 25 April 2001.

p. 150 'I thought the governors would say ...': ibid.

p. 151 It had been Hunkapiller's idea: Nicholas Wade, 'Beyond sequencing of human DNA', *New York Times*, 12 May 1998.

p. 152 'This is a very exciting development . . .': email from Mark Guyer to John Sulston and others, 10 May 1998.

p. 153 Anyone reading that article . . .: Nicholas Wade, 'Scientist's plan: map all DNA within 3 years', *New York Times*, 10 May 1998.

p. 153 'It was like asking them to walk into the sea and drown': interview with Jim Watson, 15 December 2000.

p. 154 'As I saw it, Craig wanted to own the human genome . . .': ibid.

p. 155 he followed up the presentation with an article: James Weber and Eugene Myers, 'Human whole-genome shotgun sequencing', *Genome Research*, vol. 7, 1997, pp. 401–9.

p. 155 In a closely argued response: Philip Green, 'Against a whole-genome shotgun', *Genome Research*, vol. 7, 1997, pp. 410–17.

p. 157 Officially they were to be released . . .: press release issued by the Perkin-Elmer corporation and TIGR, 9 May 1998.

p. 157 'It was a testy meeting': interview with Bob Waterston, 8 December 2000.

p. 158 'He didn't know whether I was going to belt him . . .': interview with Gerry Rubin, 11 May 2001.

p. 159 Stanley Pons and Martin Fleischmann's 'discovery' of cold fusion: see e.g. Robert Pool, 'Cold fusion: only the grin remains', *Science*, vol. 250, 1990, pp. 754–5.

p. 160 The initial press reports on Craig's company: see e.g. Meredith Wadman, 'Company aims to beat NIH human genome efforts', *Nature*, vol. 393, 1998, p. 101.

p. 160 The initial Perkin-Elmer press release . . .: Perkin-Elmer/TIGR, 'Perkin-Elmer, Dr J. Craig Venter, and TIGR announce formation of new genomics company', 9 May 1998.

p. 160 Although Craig later insisted . . .: Statement of Craig Venter before the Subcommittee on Energy and Environment, U.S. House of Representatives Committee on Science, 17 June 1998, available at <www.house.gov/science/>.

p. 161 'The public project was portrayed ...': interview with Francis Collins, 11 December 2000.

p. 163 'By the time we got to that Wednesday meeting ...': interview with Michael Morgan, 25 April 2001.

p. 167 'The atmosphere was electric': ibid.

p. 167 it was now more rather than less likely that the sequence would be completed: I said this on the basis that there would be some sort of co-operation between Celera and the HGP. This didn't happen, and so the timetable for *finished* sequence was scarcely altered.

p. 167 'It was absolutely critical, psychologically': interview with Jim Watson, 15 December 2000.

p. 167 'The talk was of healthy competition ...': editorial, 'A challenge to genetic transparency', *Nature*, vol. 393, 1998, p. 195.

p. 169 'The excitement generated by the well-orchestrated public relations campaign ...': Statement of Maynard Olson before the Subcommittee on Energy and Environment, U.S. House of Representatives Committee on Science 17 June 1998, available at <www.house.gov/science/>.

p. 170 'Any delay can result in wasted effort in research': Statement of Francis Collins, ibid.

p. 170 he and his colleagues ... had 'developed a new strategy for identifying genes': Statement of Craig Venter, ibid.

p. 170 Maynard Olson's criticisms ... were interpreted by many as sour grapes: see e.g. Kevin Davies, *The Sequence* (Weidenfeld & Nicolson, 2001), p. 153.

p. 171 It was time for Francis Collins ... to make a statement: Eliot Marshall, 'NIH to produce a "working draft" of the genome by 2001', *Science*, vol. 281, 1998, pp. 1774–5.

p. 171 'No-one else is doing this': Francis Collins, quoted in NIH news release, 'Genome project leaders announce intent to finish sequencing the human genome two years early', 14 September 1998.

p. 176 Genome Research Ltd: after various changes the board of directors had consolidated under the chairmanship of Derek Burke (previously vice-chancellor of the University of East Anglia). When Mike Dexter became director of the Wellcome Trust he took over the chairmanship himself.

p. 177 Under the heading 'Friendly fire': email from John Sulston to Francis Collins, 28 October 1998.

p. 183 It came out in *Science*: C. *elegans* sequencing consortium, 'Genome sequence of the nematode *C. elegans*: a platform for investigating biology', *Science*, vol. 282, 1998, pp. 2012–18.

p. 183 'On stage we had Francis Collins . . .': interview with Bob Horvitz, 11 December 2000.

p. 185 'In the changing environment of biology today . . .': letter from John Sulston to board of management of the Sanger Centre and Genome Research Ltd, 28 August 1998.

6 PLAYING POLITICS

p. 187 It was the day on which joint press announcements were made: see e.g. Colin McIlwain, 'World leaders heap praise on genome landmark', *Nature*, vol. 405, 2000, pp. 983–4; Andy Coghlan and Nell Boyce, 'The end of the beginning', *New Scientist*, 1 July 2000.

p. 189 'That was a turning point meeting . . .': interview with Francis Collins, 11 December 2000.

p. 189 'It was pretty clear that John's opinion . . .': ibid.

p. 190 'Bob came around, and by the end of the day John was up at the blackboard . . .': ibid.

p. 190 The accelerated timetable was announced in mid-March: Elizabeth Pennisi, 'Academic sequencers challenge Celera in a sprint to the finish', *Science*, vol. 283, 1999, pp 1822–3.

p. 190 'If met, the new date set by the consortium could allow the public venture to claim some measure of victory . . .': Nicholas Wade, *New York Times*, 16 March 1999.

p. 190 Craig . . . said that the new timetable was 'nothing to do with reality': ibid.

p. 191 'It's kind of upsetting for all of us': Glen Evans, quoted in Pennisi, 'Academic sequencers challenge Celera in a sprint to the finish'.

p. 192 'I got all that I had requested . . .': email from Bob Waterston to

John Sulston, 28 June 2001.

p. 193 'This announcement gives the impression that [we're] not needed': André Rosenthal, quoted in Pennisi, 'Academic sequencers challenge Celera in a sprint to the finish'.

p. 194 They all finally got together: Eliot Marshall, 'Sequencers endorse plan for a draft in 1 year', *Science*, vol. 284, 1999, pp 1439–40.

p. 197 'The day after we announced Celera, we set off an arms race . . .': quoted in Philip Ross, 'Gene machine', *Forbes*, 21 February 2000.

p. 198 a consortium consisting of the Wellcome Trust and ten of the big pharmaceutical companies . . .: see <http://snp.cshl.org>; also Eliot Marshall, 'Drug firms to create public database of genetic mutations', *Science*, vol. 284, 1999, p. 406; Eugene Russo and Paul Smaglik, 'Single nucleotide polymorphisms: big pharma hedges its bets', *The Scientist*, vol. 13, 19 July 1999, p. 1.

p. 200 Ian Dunham and his colleagues had also begun to look . . .: E. Dawson et al., 'A SNP map of human chromosome 22: extracting dense clusters of SNPs from the genomic sequence', *Genome Research*, vol. 11, 2001, pp. 170–8.

p. 201 The chromosome 22 sequence was published in *Nature*: Ian Dunham et al., 'The DNA sequence of human chromosome 22', *Nature*, vol. 402, 1999, pp. 489–95.

p. 202 We've since been mocked . . .: Davies, *The Sequence*, p. 194.

p. 202 'it shows that you can get very good finishing': quoted in Declan Butler, '"Finishing" success marks major genome sequencing milestone . . .', *Nature*, vol. 402, 1999, pp. 447–8.

p. 202 'We could not have done this work . . .': ibid., box headed '. . . as researchers pounce on glut of data'.

p. 203 Craig Venter's agreement with Gerry Rubin to sequence the fly genome: Elizabeth Pennisi, 'Fruit fly researchers sign pact with Celera', *Science*, vol. 283, 1999, p. 767.

p. 205 The 'public release' that Craig had promised when Celera was launched: Statement of Craig Venter before the Subcommittee on Energy and Environment, U.S. House of Representatives Committee on Science 17 June 1998, available at <www.house.gov/science/>.

p. 205 They ... announced that the sequencing [of the fly] was 'completed': Declan Butler, 'Venter's Drosophila "success" set to boost human genome efforts', *Nature*, vol. 401, 1999, p. 729.

p. 205 'other early genomes': Celera Genomics, 'Celera Genomics completes sequencing phase of Drosophila genome project', press release, 9 September 1999.

p. 205 'It has been a win–win affair': editorial, 'Human chromosome 22 and the virtues of collaboration', *Nature*, vol. 402, 1999, p. 445.

p. 205 Six weeks later Celera announced that it had sequenced 1 billion base pairs: Celera Genomics, 'Celera Genomics delivers over 1 billion base pairs of human DNA', press release, 20 October 1999.

p. 206 In November Celera invited forty fly biologists to an 'annotation jamboree': Elizabeth Pennisi, 'Ideas fly at gene-finding jamboree', *Science*, vol. 287, 2000, pp. 2182–4.

p. 206 A software tool called Ensembl: <www.ensembl.org>.

p. 207 'I did think it would be a useful thing to try to defuse this ...': interview with Eric Lander, 10 May 2001.

p. 208 No wonder Celera targeted the Sanger Centre: see e.g. Kevin Toolis, 'DNA: it's war', *Guardian*, 6 May 2000, Weekend section, pp. 9–20.

p. 209 'the Wellcome Trust is now trying to justify ...': quoted in Richard Preston, 'The genome warrior', *New Yorker*, 7 June 2000.

p. 209 'statement of principles': reproduced in Davies, *The Sequence*, p. 199.

p. 210 'We had been led to believe that they were seriously seeking some co-operation': interview with Bob Waterston, 8 December 2000.

p. 210 'It was so different from what we had been led to believe': ibid.

p. 210 Craig was already quite open about the fact that Celera was going to combine the publicly available data ...: Paul Smaglik and Declan Butler, 'Celera turns to public genome data to speed up endgame', *Nature*, vol. 403, 2000, pp. 119–20.

p. 213 In January 2000 the company announced that it had sequenced 81 per cent of the genome: Smaglik and Butler, 'Celera turns to public genome data'; Celera, 'Celera compiles DNA sequence covering 90% of the human genome', press release, 10 January 2000.

p. 216 'Celera will need to hang its sequence data on the framework produced by the public project': Smaglik and Butler, 'Celera turns to public genome data'.

p. 216 Celera described its use of our data as a 'de facto collaboration': Nicholas Wade, 'Company nears last leg of genome project', *New York Times*, 11 January 2000.

p. 217 It had a huge impact: see e.g. Eliot Marshall, 'Talks of public–private deal end in acrimony', *Science*, vol. 287, 2000, pp. 1723–4.

p. 217 Craig Venter and Tony White [called] the Trust's action 'slimy' and 'a low-life thing to do': Justin Gillis, 'Gene map alliance hopes fade', *Washington Post*, 6 March 2000, p. A04, 7 March 2000, p. E01.

p. 217 Craig even taunted us: Toolis, 'DNA: it's war'.

p. 218 The idea that it had been done for underhand motives was 'fanciful': Gillis, *Washington Post*, 7 March 2000, p. E01.

p. 218 the *Washington Post* described the genome project as 'a mud-wrestling match': ibid.

p. 218 From its place deep inside the interview . . .: see e.g. Natasha Loder, 'Rival demands sink genome alliance plans', *Nature*, vol. 404, 2000, p. 117.

p. 219 A week later Bill Clinton and Tony Blair made a joint statement: Declan Butler, 'US/UK statement on genome data prompts debate on "free access"', *Nature*, vol. 404, 2000, pp. 324–5; Bruce Alberts and Aaron Klug, 'The human genome itself must be freely available to all mankind', *Nature*, vol. 404, 2000, p. 325.

p. 219 But on the day of the statement CBS Radio News reported: Robert Langreth and Bob Davis, 'Plunge in biotech stocks linked to press briefing', *Wall Street Journal*, 16 March 2000.

p. 221 Celera announced that it had completed the sequencing of the first human genome: Nicholas Wade, 'Analysis of human genome is said to be completed', *New York Times*, 7 April 2000.

p. 221 Once again there was a most ingenious press release: Celera

Genomics, 'Celera Genomics completes sequencing of the genome from one human', press release, 6 April 2000.

p. 221 'You should not take at face value any claim . . .': quoted in Allen Dowd, 'Expert urges caution on genome discovery claims', Reuters, 10 April 2000, reported e.g. in *Wired News*.

p. 222 his department issued a partial retraction: Christopher Elser, 'Celera rival denies questioning claim Celera mapped 99% of DNA', *Bloomberg*, 11 April 2000.

p. 222 'The Sanger Centre has the support of the Wellcome Trust': interview with Francis Collins, 11 December 2000.

p. 222 Columbia University . . . had been due to hear from him at a seminar: Paul Pavlidis, personal communication.

p. 223 It was a party political issue: Nicholas Wade, 'Analysis of human genome is said to be completed', *New York Times*, 7 April 2000.

p. 223 Clinton sent a note to Neal Lane: Frederick Golden and Michael Lemonick, 'The race is over', *Time*, 3 July 2000.

p. 223 'I felt pretty uneasy about doing that . . .': interview with Francis Collins, 11 December 2000.

p. 225 Of course, the 26 June announcement was a political gesture: see e.g. editorial, 'Human genome projects: work in progress', *Nature*, vol. 405, 2000, p. 981.

7 IN THE OPEN

p. 226 in an article to appear in the journal *Science*: Craig Venter et al., 'The sequence of the human genome', *Science*, vol. 291, 2001, pp. 1304–51.

p. 227 Celera had been launched with the promise . . . : Statement of Craig Venter before the Subcommittee on Energy and Environment, US House of Representatives Committee on Science 17 June 1998, available at <www.house.gov/science/>.

p. 229 It seemed unlikely that *Nature* would bend its rules: editorial, 'Rules of genome access', *Nature*, vol. 404, 2000, p. 317.

p. 229 leading figures . . . wrote to voice their concern: Eliot Marshall,

'Storm erupts over terms for publishing Celera's sequence', *Science*, vol. 290, 2000, pp. 2042–3; Eliot Marshall, 'Sharing the glory, not the credit', *Science*, vol. 291, 2001, p. 1189.

p. 229 Kennedy responded: Marshall, 'Sharing the glory'.

p. 230 We began to think about our paper: International Human Genome Sequencing Consortium, 'Initial sequencing and analysis of the human genome', *Nature*, vol. 409, 2001, pp. 860–921.

p. 234 Michael Ashburner . . . wrote an outraged letter: Marshall, 'Storm erupts over terms for publishing Celera's sequence'.

p. 241 Despite disparaging remarks from Craig Venter: e.g. Tim Radford, 'Articles of faith lie at heart of bitter feud', *Guardian*, 12 February 2001.

p. 243 Robin's interview with Craig was to be the front-page story: Robin McKie, 'Men and women behaving badly? Don't blame DNA', *Observer*, 11 February 2001, p. 1; Laurie Cohen and Antonio Regalado, 'How the media scrambled to tell news about the human genome', *Wall Street Journal*, 13 February 2001.

p. 243 Ian Dunham's [paper] on chromosome 22: Dunham et al., 'The DNA sequence of human chromosome 22'.

p. 244 'I am sure I speak for all of us . . .': quoted in Sherri Chasin Calvo, 'At genome publication press conference spirit of unity masks tension', *GenomeWeb*, 12 February 2001.

p. 244 'Most importantly we have made this available to the world . . .': ibid.

p. 245 Aaron Zitner of the *Los Angeles Times*: '"Whole-genome shotgun" missed its mark', *Los Angeles Times*, 11 February 2001.

p. 245 A consortium consisting of three private companies, six institutes of the National Institutes of Health and the Wellcome Trust: Eliot Marshall, "Public-private project to deliver mouse genome in 6 months," *Science*, vol. 290, 2000, pp. 242-3.

p. 245 Craig Venter called (unsuccessfully) for public funding for mouse sequencing to be terminated: Nicholas Wade, 'National Cancer Institute to buy access to rival's genome data,' *New York Times,* 10 July 2001.

p. 246 Meanwhile, Bob Waterston, Eric Lander, and I wrote a brief analysis: Robert Waterston, Eric Lander and John Sulston, 'On the sequencing of the human genome,' *Proceedings of the National Academy of Sciences*, vol. 99, 2002, pp. 3712-3716.

p. 246 The following issue carried a vigorous rebuttal: Eugene Myers, Granger Sutton, Hamilton Smith, Mark Adams, and Craig Venter, 'On the sequencing and assembly of the human genome,' *Proceedings of the National Academy of Sciences*, vol. 99, 2002, pp. 4145-4146.

p. 246 ...a second commentary on our paper from Phil Green: Phil Green, 'Whole genome disassembly,' *Proceedings of the National Academy of Sciences*, vol. 99, 2002, pp. 4143-4144.

p. 248 'Each new round of press conferences . . .': Maynard Olson, 'Clone by clone by clone', *Nature*, vol. 409, 2001, pp. 816–18.

p. 255 French doctors for the first time successfully treated two babies . . .: M. Cavazzana-Calvo et al., 'Gene therapy of human severe combined immunodeficiency (SCID)-X1 disease', *Science*, vol. 288, 2000, pp. 669–72.

p. 255 So far such treatments have not led to long-term improvements: see David Weatherall, *Science and the Quiet Art* (Norton, 1996) for a discussion of the realities of transforming research into new treatments.

p. 256 the latter is currently banned: see the website of the Human Genetics Commission, which advises the U.K. government, at <www.hgc.gov.uk>.

p. 257 cases of children bringing lawsuits . . . for 'wrongful life': see John Harris, *Clones, Genes and Immortality* (Oxford University Press, 1998).

p. 258 Richard Dawkins's lovely phrase: Richard Dawkins, *The Blind Watchmaker* (Penguin, 1980).

p. 259 'It has not escaped our notice . . .': International Human Genome Sequencing Consortium, 'Initial sequencing and analysis of the human genome', *Nature*, vol. 409, 2001, pp. 860–921.

8 OUR GENOME

p. 264 a widespread feeling that science has already gone too far: see e.g. Jeremy Rifkin, *The Biotech Century* (Phoenix, 1999).

p. 268 a survey of researchers in US university labs . . .: Anna Schissel, Jon Merz and Mildred Cho, 'Survey confirms fears about licensing of genetic tests', *Nature*, vol. 402, 1999, p. 118; see also Jon Merz, Statement to the Subcommittee on Courts and Intellectual Property of the Committee on the Judiciary, U.S. House of Representatives Oversight Hearing on Gene Patents and Other Genomic Inventions, 2000, available at <www.house.gov/judiciary/merz0713.htm>.

p. 268 The guidelines on patenting genes in the US: Paul Smaglik, '. . . as US tightens up on speculative claims', *Nature*, vol. 403, 2000, p. 3; <www.uspto.gov>.

p. 270 a pragmatic step would be to make gene patents subject to compulsory licensing: Seth Shulman, 'Toward sharing the genome', *Technology Review*, September/October 2000.

p. 270 Anti-biopiracy campaigners won a significant victory . . .: David Dickson and K. S. Jayaraman, 'Aid groups back challenge to neem patents', *Nature*, vol. 377, 1996, p. 95; U. Hellerer and K. S. Jayaraman, 'Greens persuade Europe to revoke patent on neem tree', *Nature*, vol. 405, 2000, pp. 266–7.

p. 271 surely we should not be patenting whole life forms?: Quirin Schiermeier and David Dickson, 'Europe lifts patent embargo on transgenic plants and animals', *Nature*, vol. 403, 2000, p. 3.

p. 273 The editor of the *BMJ* . . .: Anonymous, 'Editor resigns from post after tobacco gift', *British Medical Journal*, 19 May 2001.

p. 273 The pressure to be accommodating is then huge: Peter Gwynn, 'Corporate collaborations: scientists can face publishing constraint', *The Scientist*, vol. 13, 24 May 1999, p. 1.

p. 274 'wealth creation and the quality of life': UK government White Paper, *Realising our Potential* (Her Majesty's Stationery Office, 1993).

p. 275 Oxfam's recent report to the government on globalization: *The White Paper on Globalisation*, Oxfam Policy Papers no. 12, December 2000.

p. 276 they were about to embark on a new campaign: 'Cut the cost', see <www.oxfam.org.uk/cutthecost/index.htm>.

INDEX

Note: **Boldface** numbers refer to pages on which the index entry is defined.

National Academy of Sciences, US 61, 63, 183, 229, 246
National Health Service (NHS) 142–3
National Human Genome Research Institute (NHGRI) 89–90, 152, 153, 161, 196, 235; as National Center for Human Genome Research 89, 107, 137, 138
National Institutes of Health (NIH), US 143, 188, 268; funding for human genome initiative 60–2, 68, 69–70, 83–90, 104, 122, 129, 136, 158, 167–8, 171, 174–5, 191; bid to patent ESTs 118; genome program 123, 245; relations with Celera 153–4, 195, 217–18; obligation to support US industry 176–7; press conference on publication of worm sequence 183; GenBank (database) 64, 95, 158, 205, 227; National Center for Biotechnology Information (NCBI) 206; National Institute of General Medical Sciences 61; National Institute for Neurological Disorders and Stroke (NINDS) 70; Scientific Advisory Council 178, 180–2
natural selection **20**, 258
Nature (journal) 11, 141, 202, 227; editorial comment 167, 203; reveals Celera's intention to use public data 216; and publication of HGP/Celera papers 228–9, 232, 233, 234, 235, 239, 241, 242, 243, 247–8
Nature Genetics (journal) 107
NCBI *see* National Institutes of Health
nematode *see Caenorhabditis elegans*
neurophysiology 16
neurotransmitters 23–4, 105
New England Biolabs (company) 150
New Scientist 245
New York Hall of Science 235
New York Times 109, 153, 154, 160, 190, 207
New Yorker 209
Newsnight (BBC TV) 220
NGOs (non-governmental organizations) 277
Nobel Prize 13, 110
Nomarski microscope 26–7, 72
Nottingham University, UK 273
nuclear magnetic resonance 18
nucleic acids *see* DNA, RNA
nucleotides **11**, 17, 74, 154
Nüsslein-Volhard, Christiane 46

Observer (newspaper) 242–4, 248
Office of Technology Assessment, US 61

Ogilvie, Bridget 91–4, 111, 114, 125, 143, 179
Olson, Maynard 57, 84, 100, 118, 123, 126–8, 137, 169, 170–1, 175, 247–8; mapping of yeast genome 45, 52–3;
Orgel, Leslie 18, 20–21
Oxfam 275–7
Oxford University 5, 18, 35

P1 artificial chromosomes (PACs) 290
patenting 87–8, 90, 110–11, 112–13, 118–19, 141, 157, 166–7, 170, 211, 220–1, 224–5, 253–4, 267–72, 276–7; European Patent Directive 268–9; European Patent Office 269, 270–1; US Patent and Trademark Office 90, 269
Paterson Institute of Cancer Research, Manchester 131
Patrinos, Ari 152, 169, 182, 223, 228
Patterson, Mark 228
Pavlidis, Paul 82, 235, 236
Paxman, Jeremy 220
PE Biosystems *see* Applied Biosystems Inc.
penicillin 39
Perkin-Elmer (PE, company) 151, 158, 160, 169, 196–7
Perutz, Max 10, 13, 14, 18, 25, 69
Peters, Keith 63
Pharmacia (company) 70, 71
phrap **117**, 241
Pons, Stanley 159
Powell, Don 220
Priess, Jim 73
Proceedings of the National Academy of Sciences, 246
proteins 6, 12–13, 40, 258, 272
proteomics 258

Quinn, Marc 3–4

Rees, Christine 101
Rees, Dai 86, 91, 125, 133, 134
Reese, Colin 17, 18, 21
RNA 17, 40, 47, 105, 119, 250
Roberts, Richard 150
Roe, Bruce 201
Rogers, Jane 123, 164, 165, 168–9, 175, 178, 197–8, 233, 238; joins Sanger Centre (sequencing) 97–8, 101, 102, 152, 185; and Sanger Centre program of work 130; on costs 132, 162; at 1999 Cold Spring Harbor meeting 194
Rosenthal, André 193
Roswell Park Cancer Institute, New York State 171–2
Royal Society, London 56, 61, 183, 229